Design Constraints for NFC Devices

Design Constraints for NFC Devices

Dominique Paret

iSTE

WILEY

First published 2016 in Great Britain and the United States by ISTE Ltd and John Wiley & Sons, Inc.

ISTE Ltd
27-37 St George's Road
London SW19 4EU
UK

www.iste.co.uk

John Wiley & Sons, Inc.
111 River Street
Hoboken, NJ 07030
USA

www.wiley.com

Library of Congress Control Number: 2016933848

British Library Cataloguing-in-Publication Data
A CIP record for this book is available from the British Library
ISBN 978-1-84821-884-0

Contents

Acknowledgements

During my long career in three disciplines – firstly as a professional and industrialist at Philips/NXP Semi-Conductors, secondly and simultaneously as a teacher at numerous engineering schools and universities, and finally, for the past ten years, as founder of dp-Consulting (a firm of consultants and independent technical experts) – I have had the opportunity to meet many experts in this field. Therefore, it is extremely difficult to thank everyone individually to whom thanks are due – that could fill an entire book and more! In addition, as NFC (Near-Field Communication) has such a vast range of applications, it would be fallacious to try to write such a book as this on the subject alone, so my heartfelt thanks go to numerous colleagues and friends:

– from NXP Semiconductors in Graz (Austria), Hamburg (Germany), Monza (Italy), and Caen (France), with whom I have worked for many years;

– customers, partners, and competitors whom I frequently meet at working meetings of the ISO and "mirror" commissions at AFNOR (France's standardization body).

In addition, in the coming chapters, I shall occasionally address some specific acknowledgements to various friends for their help.

I also devote this book to the many, many readers who have followed me over the years, through series of publications and technical reports. I owe to you many of these words of gratitude, because it is thanks to you and for you that I have been able to take my courage in both hands to squeeze as much detail out of the subject as possible, with a view to presenting some of its mysteries in as didactic a fashion as possible! In any case, thanks again to one and all for your help and your assiduousness in reading my work. If these few words of thanks have encouraged you to persevere in learning the technique and given you a vocation as a trainer and pedagogue, I would be delighted.

Once again, my true thanks to all of you for your contributions and your faithful friendship.

Preface

Why and for Whom is this Book Written?

Why

Working in the field of NFC (*Near-Field Communication*) since its very beginnings (a little over 15 years now!), I have written a great deal about this subject in the past. Many advances have been made on the basis of the operational principles, and wonderful and highly varied applications (*software*) in NFC (particularly with mobile phones). However (there is always a but), there has been relatively little detailed literature produced on the specific functions and the mysteries of application-specific integrated circuits and antennas (*hardware*) involved in NFC, and less still about the physical layers – the air interface. It is my hope that this book will go a long way towards filling that void.

Therefore, I would rather clarify the situation immediately. It is not out of pure ideology that I wish to rectify this shortcoming, but instead because it is a day-to-day reality. Designers and users often graciously overlook certain constraints, thinking "It will work fine without it"! However, communicating through the air interface, networked antennae, the various form factors of different applications and physical environments, their problems are unfortunately truths that must be taken into account, and without which nothing constructive can be achieved.

For whom

Having been a technical consultant for many years to numerous companies in navigating these mysterious waters, and in view of the thriving

NFC market, I have designed this book to serve engineers, technicians, students, and the growing number of new arrivals to this domain, to help them avoid certain pitfalls. At present, there is little information or basic technical and application training available on these subjects. Hence, today, I offer this book *to give you an initiation in the field and a specific knowledge of the constraints of design devoted to NFC devices*, and I have taken it upon myself to gather, compile, and collate a technical database made up of various generic industrial examples, concrete and highly-detailed examples of “designs for NFC devices and possibly more”, with widespread involvement from my company “dp-Consulting”.

The aim of this book is not to be a coverall in marketing, or a joyous ramble through the subject, but a technical reference offering an in-depth examination of all the technical details and all the functional and application issues pertaining to these technologies. As readers will discover over the course of the book, the field of applications and usages is enormously varied – ranging from the exchange of dematerialized data to secure payment, controlled by a pocket device (such as a mobile phone or a camera), which can also be used as a chip card or a reader, a television, in a car, etc. Thus, we have a multitude of “form factors”, environments, etc., leading to a multitude of technical problems where antennas are concerned!

Additionally, so as not to hamper readers’ understanding of the devices presented herein, I have done my utmost to make the book as clear and instructive as possible, so they can instantly see the point being made and the link between theory, technological aspects, financial factors, etc.

For now, I hope that readers will find this book enjoyable and illuminating. Above all, enjoy it, because it is for you (not for me) that I have written it! On that point, if any shadow of a doubt remains, readers are always welcome to contact me to put forward any comments, remarks, questions, etc. (provided they are constructive, of course) about the substance and the form of this book, via e-mail at dp-consulting@orange.fr.

IMPORTANT NOTES.– It must also be noted that this book is intended to complement many of my other works. Those works offer a more specific, even more detailed focus on the techniques used in contactless technology, in near fields in HF and NFC, and the implementation of those techniques. This further reading should also sate the technical curiosity of the vast majority of users. Note, also, that I have borrowed and reproduced from those books[1] a number of sections written previously, because sometimes it is very difficult not to simply rewrite the same things on the same subjects, all the more so given that those writings on the physical fundaments of NFC date from the late 19th Century (Biot, Savart, Maxwell and Laplace). Given that this field is evolving rapidly, it will inevitably be necessary to update the content of these books in three or four years' time, but in any case, until then at least readers will have the fundamental principles in place!

Warning

This book is intended to serve as an introduction to, and a detailed, instructive presentation of, the principles and constraints of NFC, and also to give people coming into this field an overall conceptual and applicational view of it.

EXTREMELY IMPORTANT NOTE.– From the very start, I wish to draw readers' attention to the important fact that, in order to give adequate coverage of the field of NFC, this book describes a great many patented technical principles, subject to the holding of licences and their associated rights (bitcodes, communication techniques, etc.). They have already been published in official professional technical texts/communications, or at conferences/ public seminars, but above all, they must be used in accordance with the legislation in force.

Dominique PARET
February 2016

1 Numerous documents and articles previously written by Dominique Paret and prepared for other technical publications have been included and adapted:

– *Applications en identification radiofréquence et cartes à puces sans contact*, vol. 1, Dunod, Paris, 2003.

– *RFID and Contactless Smart Card Applications*, Wiley, Hoboken, 2005.

– *NFC (Near Field Communication). Principes et applications de la communication en champ proche*, Dunod, Paris, 2012.

Introduction

This book is specifically devoted to the design (theoretical and practical) of antennas for "NFC devices", irrespective of their environments, sizes, and applications (readers, tags, mobile phones in card emulation mode, peer-to-peer (P2P) mode, tablets, communicating objects, TVs, photo frames, etc.).

Before going into detail about the content of the individual chapters, it is assumed that readers already have some degree of familiarity with the field of radiofrequency (RF) and know roughly what the term "NFC" covers. Therefore, after a brief recap of the basics of the technical constraints and the normative framework which must always be borne in mind, we shall carry out a detailed examination of the designs of 13.56 MHz antennas for NFC "*initiators*" and "*targets*", but also of all the problems of mutual interactions between initiators and targets which must always be taken into account in order to ensure the correct operation of an NFC setup.

Readers should take note right now that when reading this book, they should always be conscious of the problems that may be caused by the physical layer (the medium and its management). The challenge here, then, is to present this book in as appropriate and clear a way as possible to help in understanding all these issues. After much reflection and discussion, a choice had to be made in favor of an overarching, clearly-didactic presentation, so that readers can easily orientate themselves in the maze of all these laws of communication, which many of our readers will discover and employ in years to come.

The division of the table of contents gives a very clear roadmap to follow in order to gain a helpful view of the concrete design of antennas for all sorts of "NFC devices", and of the state-of-the-art in the field.

And now, dear friends, to work!

PART 1

Introduction to – and Reminders About – NFC

Introduction to Part 1

The first part of this book is divided into three chapters. In Chapter 1, the aim is to recap the operational principles of near-field communication (NFC) succinctly.

Chapters 2 and 3 describe the rigid structure of normative and regulatory constraints which apply to NFC elements, and primarily the design of their radiofrequency stages, their antennas and the connective problems directly associated therewith.

Although this first part of the book is an introduction, it encapsulates numerous technical points which are important to know in order to appreciate the application problems of today's world and of these devices.

It is now for you, the readers, to discover all this for yourselves.

1

Recap of the Principles Employed in NFC

In order to be able to look at the idea of near-field communication (NFC) in its proper context, this chapter offers a summary of the main physical principles used in NFC, which must be remembered throughout this book.

1.1. The physical fundaments of "contactless" and NFC

To aid in the understanding of the specific problems of these applications, for readers who are unfamiliar with the technical details (or who need to refresh their memories), let us give a brief reminder here about radiofrequency (RF) communications, and describe unambiguously what the term NFC covers. It is physics; not marketing. Hence, it is necessary to review some of the theoretical and physical bases of the technical side of NFC.

1.1.1. *Phenomenon of propagation and radiation*

Let us look at some of the conclusions on the physical concept of propagation of an RF electromagnetic (EM) wave in the classic case of radiation of a simple "Hertzian doublet". Later on, we will use this simple example to aid an examination of concrete cases of NFC.

1.1.2. *Classification of fields and spatial regions*

When studying the phenomenon of propagation and radiation of EM waves, the Maxwell equations tell us that the values of the radiated electric field E and magnetic field strength H depend notably on three main parameters:

– the variable time t;

– the variable distance r from the radiation source, defined in terms of $1/r$, $1/r^2$ and $1/r^3$. If one or more of these terms is negligible in comparison to the other(s), the equations may be simplified;

– the product kr, equal to: $kr = 2\pi/\lambda$.

1.1.3. *Spatial regions*

In the case of a very small antenna (whose sphere equivalent to the radiation source has a very small diameter), it is customary to define three spatial regions known, respectively, as the near-field, intermediary-field and far-field zones depending on the possible values of r in relation to the value of the wavelength λ at which the system operates – related to the value of r is greater or lesser than:

$$\frac{1}{k} = \frac{\lambda}{2\pi}$$

1.1.4. *Far field: $r >> \lambda/2\pi$ (Fraunhofer zone)*

In the far field zone, the terms in $1/r^2$ and $1/r^3$ present in the Maxwell equations become negligible, and the values of the fields E and H shrink in $1/r$. This occurs when the distance r is greater than a few $\lambda/2\pi$. What happens at this point is that we move from a region of quasi-stationary waves to a region of wave propagation, and thus to a regime of radiated field. The electrical field E is predominant, and there can no longer be any magnetic coupling. Generally, in this zone, devices work by using the radiating field E, so use a dipole antenna.

1.1.5. *Intermediary field: r approximately equal to λ (Fresnel zone)*

In this spatial region, all the terms in $1/r$, $1/r^2$ and $1/r^3$ are equally preserved. Here, we are in the presence of a radiating near field, where the Fresnel component exists when the antenna's primary dimension is larger than the transmission wavelength λ but the distance *r* is approximately equal to λ.

1.1.6. *Near field: r << λ/2π (Rayleigh zone) and the origin of NFC*

In the spatial region where r << λ/2π – *irrespective of the frequency at which the system operates* – i.e. often in the immediate or near surroundings of the antenna, the values of the terms in $1/r^2$ or $1/r^3$ are far greater than those of the terms in $1/r$, and the fields (*E*, *H*) decrease very quickly. This near-field zone – often called the Rayleigh zone–essentially corresponds to the region of a "(more or less) coupled transformer" (where the received power P_r is basically equal to the transmitted power P_t, unlike the effect of free space propagation, where $P_r << P_t$ (far field)).

When the transmission takes place very near to the antenna, the electrical field is generated by the antenna's own potential: this is the near-field domain. At this distance, we see a regime of "quasi-stationary" waves, and it is then possible to establish magnetic or electrical coupling between a tag and the source.

Generally, in this zone, devices work by using the magnetic field *H* by inductive coupling, i.e. use a loop antenna, which is the case with NFC.

To summarize:

$$\text{near field (magnetic coupling)} < \frac{\lambda}{2\pi} < \text{far field (wave propagation)}$$

1.1.7. *Remarks on contactless, RFID and NFC applications*

These general remarks are of fundamental importance for remote-powered systems (when the transmitted wave provides enough energy for the tag to function). For existing semi-conductor technologies and the regulations in force, as a preliminary point, it is always interesting to know the regions of the near-field and far-field zones for which the contactless/RFID/NFC application is intended to function. With this in mind, let us look at Table 1.1, which indicates the relations between the values of the frequencies and the distances λ/2π associated therewith.

	Frequencies	Wavelengths	$\frac{\lambda}{2\pi}$	Near-field zone for NFC application	Start of far-field region > ~ 3 λ
LF	125 kHz	2,400 m	382 m	Always	1,100 m
HF	13.56 MHz	22.12 m	3.52 m	Always	11 m
UHF	433 MHz	69.2 cm	11.02 cm	0 to ~ 12 cm	24 cm
	866 MHz	34.6 cm	5.51 cm	0 to ~ 6 cm	18 cm
	915 MHz	32.8 cm	5.22 cm	0 to ~ 5.5 cm	17 cm
	2.450 GHz	12.2 cm	1.94 cm	0 to ~ 2 cm	6 cm
SHF	5.890 GHz	5.1 cm	0.8 cm	0 to ~ 1 cm	3 cm

Table 1.1. *Relations between the values of frequencies and the distances λ/2π associated therewith*

As this table illustrates, in ultra high frequency (UHF) and super high frequency (SHF), most of the time the operating distances required/possible for RFID applications (for example, reading of labels on pallets of goods) are significant (several meters). Given that these distances are always much greater than the value of λ/2π, the UHF tags used for RFID work in the "far field", and therefore must use principles other than that of magnetic coupling to communicate with the base station – for instance, they could exploit the phenomenon of propagation of EM types and backscattering of the received waves.

However, in the low frequency (LF) and high frequency (HF) bands, the operational distances required/possible for the applications – for technological reasons, because of the regulations in force, the possibilities of remote powering and cost factors – are always much smaller than the value of $\lambda/2\pi$. In this case, the devices operate and communicate in the "near field".

Since their first applications in LF and HF (particularly at 13.56 MHz), contactless chip cards and RFID tags communicating at 125 kHz and/or 13.56 MHz operate in the "near field", and therefore (to paraphrase Molière, in *The Bourgeois Gentleman)* have been operating NFC all their lives and did not even know it. Thus, in terms of its principle, in LF and HF, NFC is not intrinsically any great leap forward. Hence, the applications are based on principles which primarily involve inductive loops and inductive/magnetic coupling.

1.2. The concept of NFC

Let us now examine the physics of the applicational concept of NFC.

In 1999, the two main leaders in the semi-conductorer market – Philips Semiconductors (a Dutch company, which has since become NXP) after having bought out the Austrian company Mikron); and Japan's Sony (the respective creators of the patented designs for contactless chip cards MiFare™ (based on the international norm ISO 14443 part A) and FeliCa™ (conforming to the Japanese norm JIS X6319-4)) – came together to co-invent the concept of NFC for multiple, generic and varied applications for digital data, and not to supplant or compete with chip-card applications as many people believe. Hence, NFC was born around the year 2000.

To conclude this brief introduction, if readers still believe that NFC is synonymous only with mobile telephony, they are wrong, as proved by the following text from "Scope" (late 2002), an introduction to the norm ECMA 340 NFC IP1, which was too become ISO 18092:

> "Near Field Communication Interface and Protocol (NFCIP-1) defines communication modes using inductive coupled devices operating at the center frequency of 13.56 MHz for interconnection of computer peripherals. It also defines both the active and the passive communication modes of NFCIP-1 to

realize a communication network using NFC devices for networked products and also for consumer equipment".

1.2.1. *Biot–Savart law*

Having clearly defined NFC and demonstrated its position in relation to other technologies, let us now give a recap of the physical theoretical minimum of what needs to be recalled to gain a detailed view of what is flowing through the coils and loops of antennas and what we are talking about throughout this book. We must remember that an NFC device only works if layer 1 – the *physical layer* in the OSI model – works properly. This involves:

– the creation of a magnetic field H;

– the evolution of that field with distance.

The law, known as the Biot–Laplace and Biot–Savart law, establishes the mathematical relation which exists between a current I circulating in an element of an electric circuit of length dl and the intensity of the magnetic induction B as a function of the distance x existing between the measured point (of unit vector u) and the oriented element dl producing the field, according to the formula:

$$d\,H = \frac{(I\,dl)\wedge u}{4\ \pi x^2}$$

so that $B = \mu \times H = \dfrac{(\mu\, I)}{4\,\Pi} \oint \dfrac{dl \wedge u}{dx^2}$

where $\oint$ is the curvilinear integral calculated on the length l and $\wedge$ is the notation of the vector product.

1.2.2. *Field H at a point on the axis of a circular antenna*

In the particular case of a flat circular antenna, with radius r, at a point P situated at a distance d along the axis of the antenna (Figure 1.1).

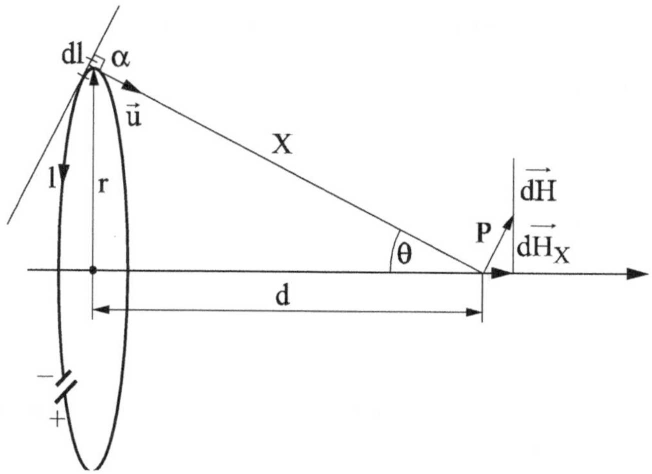

Figure 1.1. *Field H at a point on the axis of a circular antenna*

We can write:

$$dH = (I\ dl\ ) \wedge u)/4\pi\ x^2$$

and hence:

$$dH = (I\ dl\ \sin\alpha)/4\pi\ x^2$$

As, in this case:

$$\alpha = \pi/2 \text{ and } \sin\alpha = 1$$

so:

$$dH = I\ dl/4\pi\ x^2$$

On the axis of the coil, the component dH_x of dH is equal to:

$$dH_x = I\ dl\ \sin\theta/4\pi\ x^2 \text{ and } x = r/\sin\theta$$

and hence:

$$dH_x = I\ \sin^3\theta\ dl/4\pi\ r^2$$

In addition, at a distance *d* along the axis of the coil:

$$x = \sqrt{(r^2 + d^2)} \text{ and } \sin^3 \theta = r^3/x^3$$

so:

$$\sin^3 \theta = r^3/(r2 + d2)^{3/2}$$

which gives us:

$$dH_x = [(I/4\pi) \times (r/(r^2 + d^2)^{3/2})] \times dl$$

Thus, definitively, by now integrating that simple integral in *dl* along the whole length of the circular spire of length 2πr:

$$H_x = [(I/4\pi) \text{ x } (r/(r^2 + d^2)^{3/2}] \int_0^{2\pi r} dl$$

we obtain:

$$H(d, r) = \frac{I\, r^2}{2(r^2 + d^2)^{3/2}}$$

Furthermore, B = μ H(d, r):

$$B(d, r) = \mu \text{ x } \frac{r^2}{2\,[(r^2 + d^2)^{3/2}]} I$$

EXAMPLE 1.1.– In the air, with $\mu = \mu_0 = 4 \times 3.14 \times 10^{-7}$:

– if H = 1 A/m then $B = \mu_0 \times H = 1.256$ μT;

– if B = 1 μT then $H = B/\mu_0 = 0.796$ A/m.

It should be noted that for d = 0 and a coil of N spires, we see the same, well-known formula for the induction B in the center of a flat circular coil:

$$B(0, r) = \mu \frac{N\, I}{2\, r} = \mu \; H(0, r)$$

1.2.3. *Decrease in the field H as a function of "d"*

By setting:

– *d* as the distance between the center of the antenna and the measuring point on the axis;

– *r* as the radius of the antenna;

– a = d/r and d = (a r).

The above relation between H_0 (in the center of the base station antenna) and H_d becomes:

$$H(a, r) = \frac{1}{[(1 + a^2)^{3/2}]} \times H(0, r)$$

The ratio H(a, r)/H(0, r) depending on the values of *r* and *d*, or indeed on the values of the ratio *a* = *d/r*, are shown in Table 1.2 (for a given value of r, the table indicates the values of the corrective factor "$(1+a^2)^{3/2}$" and its inverse as a function of the ratio *a* = *d/r*).

d	a	$H(0,r)/H(a, r) = (1 + a^2)^{3/2}$	H(a, r)/H(0,r)	$= 1/(1 + a^2)^{3/2}$
0	0	√ 1	= 1	= 1
0.5 r	0.5	√ 1.953	= 1.397	= 0.716
r	1	√ 8	= 2.828	= 0.354
1.33 r	1.33	√ 21.34	= 4.62	= 0.216
1.414 r	1.414	√ 27	= 5.196	= 0.192
1.5 r	1.5	√ 34.33	= 5.86	= 0.170
1.66 r	1.66	√ 53.8	= 7.34	= 0.136
2 r	2	√ 125	= 11.8	= 0.0847
3 r	3	√ 1000	= 31.62	= 0.0316
4 r	4	√ 4913	= 70	= 0.0143
4.5 r	4.5	√ 9595	= 98	= 0.0101
5 r	5	√ 17576	= 132	= 0.0076

Table 1.2. *Variations of the ratio H(a, r)/H(0, r) depending on the values of "r" and "d"*

NOTE.– We will frequently refer back to this table as it is very relevant throughout the book.

Figure 1.2 reflects the results given in this table.

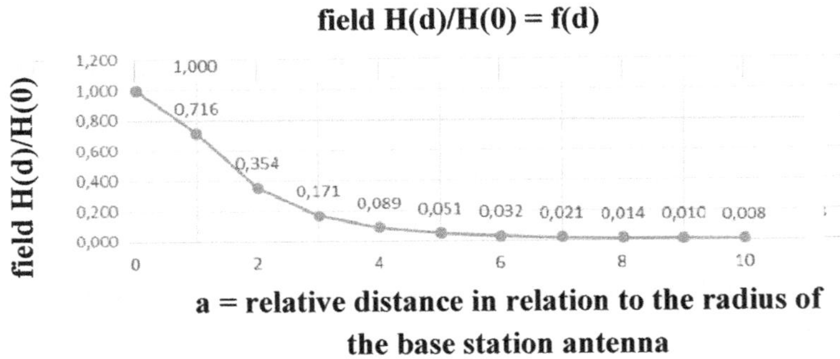

Figure 1.2. *Curve of the variations in the ratio H(d)/H(0) at "a"*

EXAMPLE 1.2.– For an antenna of an initiator/reader (proximity coupler device (PCD)) ISO 10 373-6, so of radius $r = 7.5$ cm (diameter = 15 cm) we switch from a 7.5 A/m field at 0 cm to a 1.5 A/m field (i.e. a ratio of 5), when the ratio of the distance between the antenna of the PCD (reader) and the proximity integrated circuit card (PICC) (target) is around (1.4 × the radius of the PCD), which is a distance of 1.4 × 7.5 = 10.5 cm. On principle, a PCD yielding a magnetic field H(0) of 7.5 A/m can read all ISO PICCs (because it needs to be sensitive to at least 1.5 A/m minimum) from 0 cm to around 10.5 cm.

1.2.4. *Field H at a point on the axis of a rectangular antenna*

Similarly, in the case of a flat, rectangular antenna with sides a and b, with N spires, at a point M situated at a distance x on the axis of the antenna (Figure 1.3).

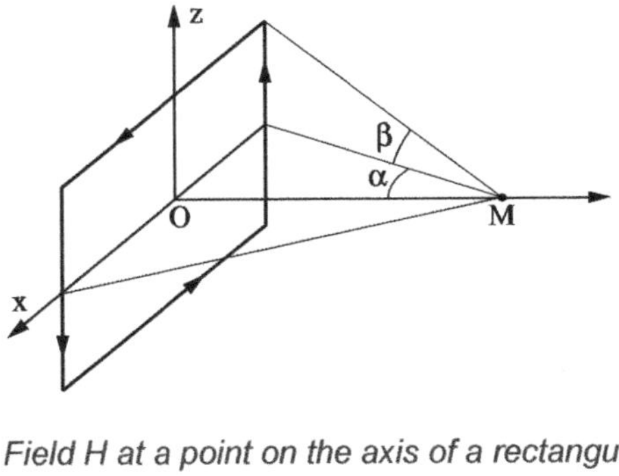

Figure 1.3. *Field H at a point on the axis of a rectangular antenna*

After a few lengthy calculations, we obtain:

$$H=\frac{N\times I\times a\times b}{4\times\pi\times\sqrt{(\frac{a}{2})^2+(\frac{b}{2})^2+x^2}}\times\left(\frac{1}{(\frac{a}{2})^2+x^2}+\frac{1}{(\frac{b}{2})^2+x^2}\right)$$

After these few physico-theoretical examples, let us now examine the detailed calculations of NFC initiator and target antennas.

2

Normative Constraints of NFC

2.1. Introduction

2.1.1. *Normative, regulatory and NFC market constraints*

On paper, in order to transform any system into an NFC system, we only need an integrated circuit, several miniature passive R, L, C components and an antenna. That is the theory at least! In reality, this migration is much more complex as, owing to the radio electric properties of the new element, the antenna, and the immense applicative field of the NFC, there are large number of constraints that must be dealt with, respected and satisfied to make the final solution viable and reliable!

Before getting into the detail of the content of this book, it is always good and instructive to know what is actually allowed, and what is not! To do this, we will begin by carrying out a detailed inventory of the main generic constraints that are usually found with NFC-type applications and mainly their direct implications on the design of their antennas. The two following chapters therefore define this context, the main constraints and the structural problems linked to the intrinsic content of the NFC protocols and those related to them, which it is necessary to satisfy to be able to use the name – in the legal sense (or otherwise risk attacks deceiving the public) – “conforms to norm NFC ISO 18 092 or 21 481 or NFC Forum standard”.

As you will discover in the next two chapters, the main constraints are of various orders, normative, regulatory and markets. But before that, let us provide a little foreword.

2.1.2. *A little bit of vocabulary*

As you will notice throughout this book, it often helps to go through a little bit of vocabulary. We are far from possessing an innate knowledge of all the terms and vocabulary, but more than 35 years as a member of the ISO makes it much easier to understand! Many application or sector branches use all types contactless systems, each given a more specific name more or less according to (or not) its own "base stations" or contactless "tags". Before going into detail in this chapter, given that the norms and/or standards have a very specific vocabulary, here is again several terms to help us understand. Table 2.1 gives several examples.

ISO group		Transmitter	Responder
		Base station	Target/transponder
		Reader	Card
		Modem	
		Coupler	Badge
SC 17 WG8	Proximity cards and personal devices	PCD Proximity coupler device	PICC Proximity integrated circuit card
	Vicinity cards and personal devices	VCD Vicinity coupler device	VICC Vicinity integrated circuit card
SC 31 WG 4	Item management/RFID	Interrogator	Tag
SC 06	NFC	Initiator	Target
		etc.	etc.

Table 2.1. *ISO terminology for the main contactless transmitters and responders*

In this book, which deals exclusively with NFC (originally developed by the ECMA in Switzerland, and then taken up again by Sub-Committee SC 06 at the ISO in 2000), we shall employ only the official ISO terms: "*initiator*" and "*target*". Thus, from this point on, having looked briefly at the principle above, all other terms will be proscribed (or almost)!

Before we begin, here are a few more definitions!

2.1.3. *Norm*

Trying to be as brief as possible, a “norm” consists of documents or rules that have been proposed, discussed, ruled on with consensus and voted for industrialists brought together under the umbrella of national (AFNOR, DIN, BSI, ANSI, etc.) or international (ISO, ECMA, CEN, etc.) bodies. The application of a norm is not automatically compulsory according to laws, orders and/or decrees, but that the industrialists apply together, voluntarily in order to ensure that their products will work together and so increase their markets. For example, technical specifications NFC IP1 and P2, called “open” by the NFC, were defined by the ISO in 2002.

2.1.4. *Standard*

For its part, a “standard” consists of documents or rules that have been proposed, discussed or ruled on (with or without consensus) by industrialists brought together within private groups (NFC Forum, etc.) and for which the application of the standard is left to the goodwill of this group and which some industrialists apply in order to increase their market. For example, technical specifications called “proprietary” by the NFC began to become available from 2006.

2.2. Normative constraints

When designing an NFC system and the associated antennas, the technical and protocol constraints needing to be respected are, obviously, the legislational and physical constraints pertaining to the “low layers” 1 and 2 of the OSI model (which are respectively the physical and data link/medium access layers), without which the whole setup could never work.

The forms (appearances, amplitudes, etc.) of NFC signals that must be respected are described in detail in the *international standards* ISO 18092 NFC IP1 and ISO 21481 NFC IP2 – the lone true standards – which draw extensively on the contactless proximity chip-card standards ISO 14443 A & B (including numerous classes of antennas – 1-6) and on Japanese standard JIS X6319-4 for the patented product FeliCa, and finally, those surrounding ISO 15693. In addition to these, there may be *proprietary- and/or market-sector-specific standards* such as (mainly) NFC Forum, EMVCo, CEN, etc. where the specific operational application characteristics of distances and volumes (in cm^3) involve antennas very frequently with different adaptations. Finally, to add the finishing touches, you sprinkle everything with application layers (OSI 6 & 7 layers) from Android and HCE from Google, Apple, Samsung... or others!

Before quickly revisiting these technological legacies and other new points specific to norms/standards that we need to design a specific antenna and conforming products, carrying out again a reminder of the basic principles and once more hammer home the specific vocabulary.

2.2.1. *Uplink from initiator to targets*

In this book the exchanges taking place between initiators and targets are defined, once and for all, as follows:

– "from the initiator to the target" known as "uplink";

– "from the target to the initiator" known as "downlink".

In order to avoid any comprehension problems, note now that, regardless of the intelligence built into the target, it only functions on the basis of "commands" sent by the initiator.

2.2.1.1. *Forward link – from the initiator to the target*

The forward link/communication is, in principle, triggered and assured by the initiator which emits a RF wave. This is thus equipped with an emitter, a "*TRANSmitter*". Its job is to:

– if possible, manage the transport of the energy to the target to enable the power supply and ensure it functions;

– act as a means of sending data from the initiator to the target.

NOTE.– In the case of systems working in the "passive" mode (see the definition later in this chapter), the carrier wave emitted by the initiator also has the job of ensuring the presence of a physical support during the return link for communication from the target to the initiator.

The initiator also includes a "*reCEIVER*" to pick up and interpret communication in the other direction.

Therefore, the initiator is a *trans… ceiver*; a *transceiver*.

In the forward direction, the initiator must be understood by the target through a numerical (binary) code to ensure the communication protocol and through a carrier frequency modulation system which must not affect, or affect only a little, the quality of a hypothetical remote power supply. In many of the usual cases of commerce applications, the amount of energy transferred is enough to ensure the remote supply of the "tag/poster"-type targets all the while ensuring the forward link of communication from the initiator to the target. Let us start by examining this potential energy supply.

2.2.1.1.1. Supplied energy and the concept of supplying the target

In order to prevent confusion, two potential scenarios are officially defined by the ISO 19762-3:

– either the energy transmitted by the RF wave from the initiator (see above) is able to provide the necessary power to the target, and in this case the target is "*remote-powered*" or "*batteryless*";

– or the energy transmitted by the RF wave emitted from the initiator is unable to remotely power the target and of course, it is important to do something about that. In this case, the target is "*battery-assisted*".

2.2.1.1.2. Remote-powered or batteryless targets

For economic reasons, as well as issues relating to weight, volume, bulk, etc., many applications do not want the targets to include additional specific power supply sources on board. The energy (and consequently the continuous power supply) that the target requires must thus be supplied to it (via antenna) through the electromagnetic field radiated from the RF wave coming from the initiator. This supposes of course that the supply is sufficient or greater than what is needed to ensure the correct functioning of the target throughout the extent of its applications.

2.2.1.1.3. Battery-assisted targets

For multiple reasons (desired working distances, technologies used, regulations in force, hostile environment, etc.) the energy transmitted by the wave emitted by the initiator is not enough to remotely power the target. To compensate for this, it is thus necessary to have, on-board the target, an independent local source of energy, a fuel cell, a battery, an accumulator, etc., which is called "on-board power supply". We thus say that the target is battery-assisted. This also enables us to be able to have more important functional possibilities available on board the target (the case of mobile phones and tablets where it is necessary to supply local electronics, the screen, the SIM card, etc, or in cases where the communication demands are higher).

On the other hand, the energy lifetimes of these targets are directly linked to those of their fuel cells, batteries, etc., with or without recharging devices and their own consumptions. In order to solve some of these problems, we usually have monitoring systems on board to enable us to prolong the lifetime of the battery and as a result that of the targets...and design other types of antennas.

NOTE.– Very frequently – too often, in fact, and *incorrectly* – remote-powered targets are said to be "passive", whilst battery-assisted ones are said to be "active". This makes absolutely no sense (see explanations later on for details).

Nevertheless, it is impossible to put the point across more clearly. Here again, it is important to try to choose the right word.

2.2.1.2. *Return link: from target to initiator*

Here again we find the ambiguous terms "passive" and "active" concerning the return link from the target to the initiator! Once again, far from wanting to profess that we alone are correct, the highly experienced technical people who work in this field day in day out have with great precision defined a certain number of terms relating to all of this (ISO 19762-3).

2.2.1.2.1. Active and passive targets

Independently of the type of power supply to the target (remote-powered or battery-assisted), a device must have an electronic means of communication to perform downlink from the target to the initiator – known as the "return link". Downlink can take place in a number of different ways depending on the principles which are used:

"Passive" targets

In the first case, having a physical support of the forward wave, generally unmodulated and supplied by the initiator, the target succeeds in making itself understood and responding to the initiator in "half-duplex" mode without the help of an on-board emitting function. We thus call this type of target "passive". Nothing prevents it being with remotely supplied by the incident wave (and thus battery less), or being battery-assisted!

"Active" targets

In the second case, the target is not capable (for x reasons, distance too great, lack of sensitivity of the initiator demodulators, etc.) of making itself understood to the initiator without the presence of an actual transmitter on board, and so one is installed! In this case and always independently of the type of power supply, due to the presence of the transmitter serving as a return link we call the target "active".

In summary, completely independently of the way in which the target is powered, the adjective "passive" defines the fact that the return communication link – from the target to the initiator – is carried out without the help of an RF transmitter.

On the other hand – and always independently of the way in which the target is powered – if it includes an on-board transmitter to respond to the initiator, the target is "active".

It is important not to confuse the power/energy transfer system with the principles of upward and downward communication.

2.2.1.2.2. Return communication from targets to the initiator

Load modulation

In the case of return links from passive targets to the initiator (which is the most common case, currently 90%), it must be recalled that the target does not have an on-board transmitter but that it must be an element capable of making itself understood by the initiator. In order to achieve downlink, the initiator provides a physical support in the form of a sustained, non-modulated "carrier" frequency, and allows the target to act however it deems fit depending on its own way of working, in order to communicate with the initiator by modulating its electrical characteristics. For this purpose, there are two distantly related modulation techniques:

– the first, based on a principle of "modulation of impedance (resistance and/or reactance) of the target antenna's load" – known as passive load modulation (PLM) is used by most targets available on the market;

– the second, which is more recent, is known as active load modulation (ALM), and is described in detail in the following sections.

In view of the physical consequences which stem from these two forms of near-field modulation, we typically speak of "retro modulation" by "magnetic coupling".

Passive load modulation (PLM)

The target, at its own pace, uses a specific all or nothing binary code (OOK – *on-off keying*) to modulate all or part of the equivalent electric charge (resistance or capacitance) that it shows at the terminals of its own antenna (see Figure 2.1).

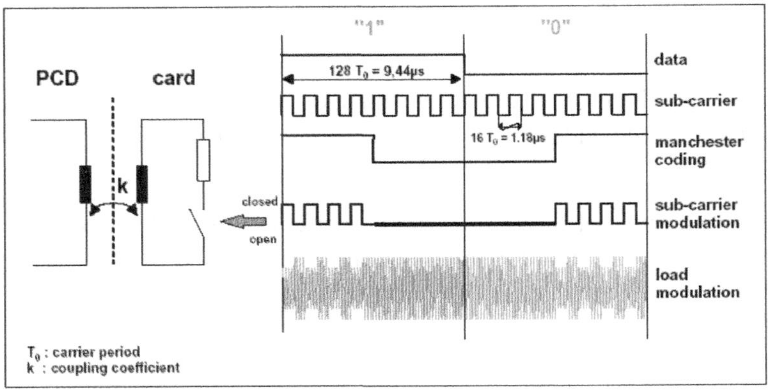

Figure 2.1. *Passive load modulation*

By modulating its load, it modifies the energy consumption that it shows and, owing to the magnetic coupling that exists between the target and initiator, this variation slightly modifies the value of the current circulating in the circuit of the initiator's antenna. Let us recall that in this case, the target does not include any actual transmitting function. The system works according to the "*TRANSmitter*" (initiator) and "*resPONDER*" (target) mode and the whole thing is in "*transponder*" mode. The communication from the target to the initiator is thus based on the principle of passivity of the return link from the target element and this is thus an appropriate antenna!

Active load modulation (ALM)

In this case, during the target response phase, two cases can be produced. Pay attention, it is not that straightforward!

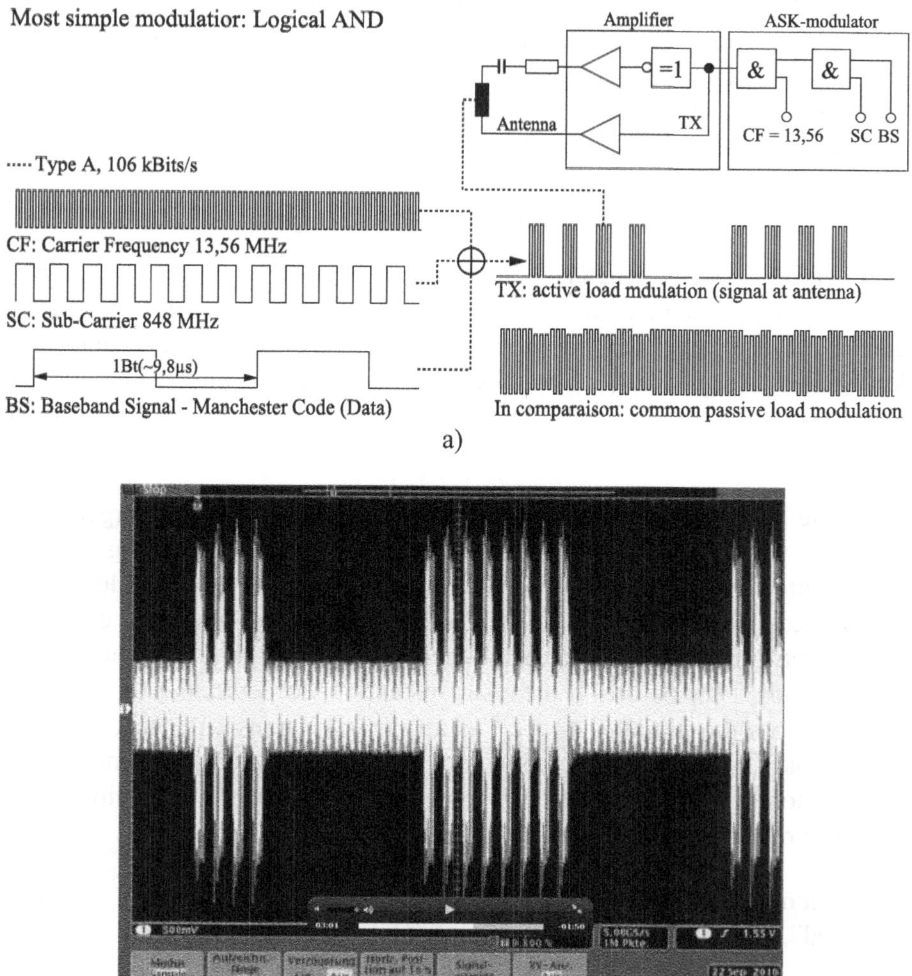

a)

b)

Figure 2.2. *Active load modulation*

Active load modulation – ALM in NFC passive mode

As previously, the NFC is in "NFC passive mode", i.e. the initiator, provides a carrier wave serving as a means of return communication during the downlink. Unfortunately, it is possible that the signal level sent by the target by PLM during the return communication is too weak to be processed by the initiator. To compensate for this, always independently of the way in which the target is powered (battery less or battery-assisted), the target is provided with a local low-power (*mini*) transmitter so that *for this period of time only* the signals returned to the initiator are "boosted"... thus with a new appropriate antenna! Due to the momentary presence of this mini-transmitter on the return leg, this system is ALM, which, over time (according to the applications), tends to be partially replace PLM (see Figures 2.2(a) and (b)).

Active load modulation – ALM in NFC active mode

In certain systems (such as NFC devices in "peer-to-peer" mode), during the downlink phase in a *half-duplex* system, the initiator no longer provides a carrier to serve as a support for the return signals. In this particular case, known as *active-mode NFC*, in order to communicate with the initiator, the target transmits its own wave, calibrated to the same frequency as that of the carrier, and becomes an "active" NFC device.

In the latter two cases presented above, the downlink of the target to the initiator works during the retro-modulation phase according to the "transmitter" mode. The return communication from the retro modulation is thus based on the principle of "activity", and so is sometimes called an "active retro-modulation" target. Of course, to achieve this, each target must have on board (as before) a means of reception (to detect the forward link coming from the initiator) as well as a (mini) RF transmitter to ensure the return link. Of course these types of operations re more complex and require a little more energy for the target, sometimes (or even often) leading to the presence of a battery on-board the target!

In conclusion, Table 2.2 sums up the "passive", "active", "remote-powered", "battery-assisted", etc., functions.

"Supply" versus "tag to interrogator communication"		
	Communication from tag to interrogator	
	Via load modulation	Via transmitter
Supply		
No battery on board	Passive Battery less	Active Battery less
Battery on board	Passive Battery-assisted	Active Battery-assisted

Table 2.2. *"Passive", "active", "remote-powered" and "battery-assisted" functions*

Now that this little review of the vocabulary has been carried out, we can focus purely on technique, examining how passive and active targets work.

2.2.2. *"Contactless" normative constraints on NFC device antennae*

As we wish this work to focus on design constraints for all types of "NFC devices" and for newcomers in this new world to have everything necessary to understanding and using it to hand, we have grouped here the principle technical points of the norms for contactless smart cards, Proximity ISO 14 443 and JIS X6319-4, and those of the contactless smart cards, Vicinity ISO 15 693, all of which are considered to be the "legacy" to which additional points have then come to be added to create the "NFC" design and norm.

For greater clarity, in the following sections, we have divided the presentation of these legacies into "uplinks" and "downlinks" and we will now examine existing technology for up and down links where they affect bit encodings, sub-bit encoding, carrier modulations, modulation rates, and systems for managing collisions with a view to singularizing systems.

2.2.2.1. *Uplink: initiator to target*

Let us first examine bit encodings.

2.2.2.1.1. Bit encodings and the uplink

Several encodings are used.

Modified-Miller (MM)

The bit encoding "single Miller" has a mid-bit transition for a "1", no mid-bit transition for "0", "and a transition at the end of the high/low bit transition for a "0" if the following bit is also a "0". In order to encode a bit according to the "modified-Miller" method (used in ISO 14 443 – 2 type A and in NFC IP1 – ISO 18 092 & IP2 – ISO 21 481), a pulse of short duration called a "pause" is created at each transition of the Miller coding and it is necessary to use different sequences (see Figure 2.3):

– sequence A, after a time t_1, a pause of duration "tp" should be produced;

– sequence B, during the duration of the bit no modulation should be produced;

– sequence C, at the start of the bit, a pause of duration tp should be produced.

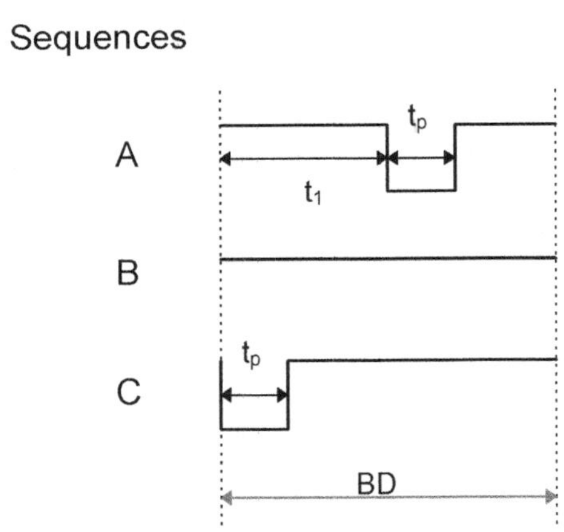

Figure 2.3. *The different sequences of modified-Miller coding*

These different sequences are used to code the three following logic states: "1" logic, "0" logic, "no information".

The "pause duration" is one of a system's major constraints as it is one of the key parameters and is fundamental for designing an antenna as it defines its quality coefficient "Q".

No return to zero (NRZ)

As its name indicates, in this encoding the level representation of the bit's value remains constant throughout its duration, (used in ISO 14 443–2 type B and in NFC IP1 & IP2).

Manchester

To encode a bit in Manchester (often called "biphasic code") it is necessary to carry out an exclusive "OU" ("XOR") between the clock signal and the data signal (used in JIS X6319-4 standard–FeliCa and in NFC IP1 & IP2):

– the"0" always has the same shape as a negative mid bit transition;

– the"1" always has the same shape as a positive mid bit transition.

Bit encoding–vicinity

Two bit encodings are used for the uplink known as vicinity in ISO 15 693, revised by the ISO in NFC IP2, and in the NFC Forum under the name of NFC-V technology.

Position encoding "1 of 256"

A byte's decimal value (from 0 to 255) is represented by the temporal position of the carrier's "pause" signal in one of the 256 possible time slots defining the byte's total duration (this pause is also one of the constraints and one of the key parameters of antenna design as if defines its own "Q"). Figure 2.4 illustrates the encoding 1 of 256 for 10% ASK modulation.

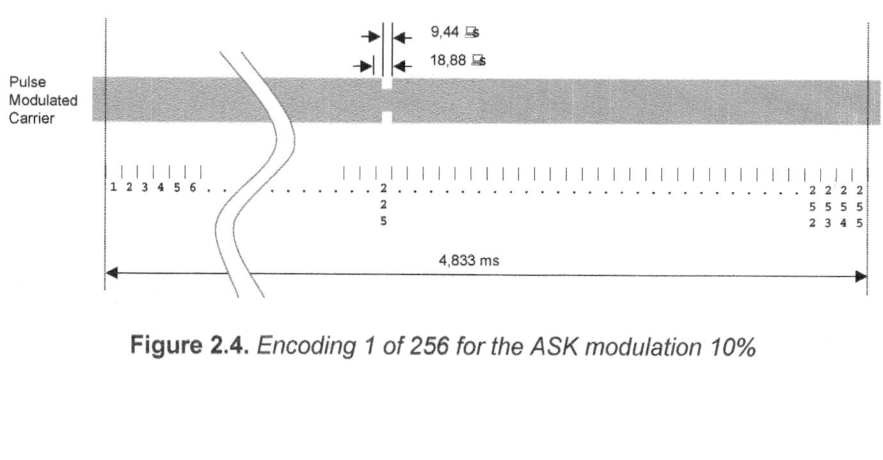

Figure 2.4. *Encoding 1 of 256 for the ASK modulation 10%*

Position encoding "1 of 4"

The pulse-position modulation known as "1 of 4" enables a two-bit symbol to be encoded in a given time slot. In the interval of time that the byte lasts, pulses representing bit pairs are successively transmitted. The unit thus constructed enables the byte transmitted to be described completely. Figure 2.5 shows this principle adopted according to the time slot.

The duration of the "pause" is one of a system's major constraints as it is one of the key parameters for designing an antenna as it is fundamental for defining its quality coefficient "Q".

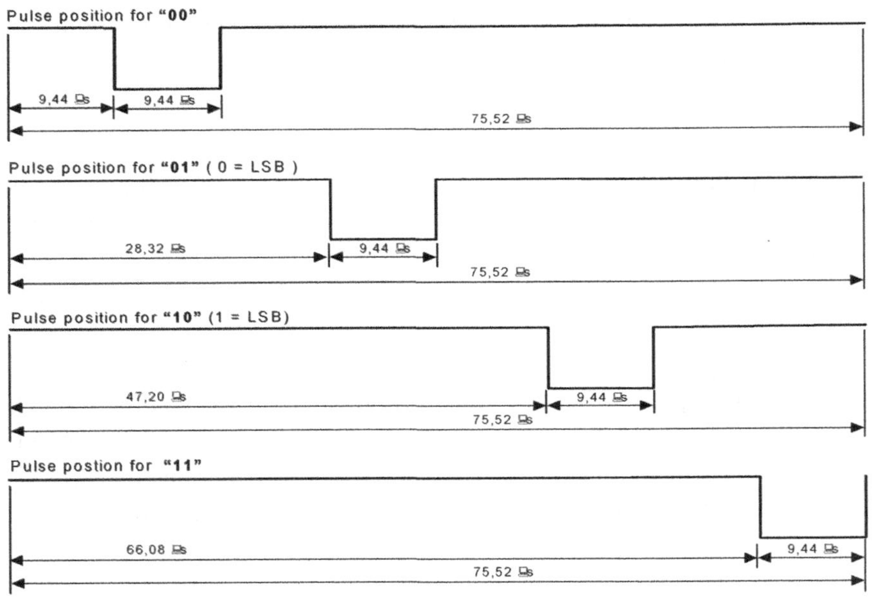

Figure 2.5. *Position encoding "1 of 4"*

2.2.2.1.2. Bit rate

Again, we can count several communication rates in existence according to the norms and standards used, which are one of the key parameters for designing an antenna as they define the bandwidths "Bp" and therefore "Q":

– ISO 14 433;

- normal rate: 106 kits/s (this rate is used during collision initialization and management),

- high rate: 212, 424, 848 kbits/s,
- very high rate: from 1.6 to 27 Mbits/s,
- JIS X6319-4: rate 212, 424 kits/s.

2.2.2.1.3. Carrier modulation and uplink

Always in view of the constraints when designing NFC antennae, and within the strict context of contactless applications, for the uplink, initiator to target, it is possible to use several types of carrier modulation to transmit a bit stream, hence the value of the antennae bandwidths Bp and Q!

Amplitude shift keying modulation

Whatever the communication direction, the NFC uses amplitude shift keying (ASK) carrier modulation (m). The value "m" defined the "modulation index–m%". It is defined by the following equation:

$$m = \frac{U\,\max - U\,\min}{U\,\max + U\,\min} * 100$$

with:

– U_{max} amplitude/tension present before modulation;

– U_{min} amplitude/tension present during modulation.

100% ASK modulation

This modulation is used in ISO 14 443-2 type A and is linked to a modified-Miller bit encoding. The shape of the signal obtained after 100% ASK modulation resembles the curve shown in Figure 2.6.

Each time the carrier is modulated (to 100%) a "pause" is observed in the carrier. For many reasons (the antennae's quality coefficient, the carrier's disconnection, connection time, etc.) linked to implementation of this technique, this pause is rarely entirely perfect.

Figure 2.7 shows the way in which this pause is usually characterized. (This is one of the very strong constraints to observe and one of the key parameters for designing an antenna as it defines its "Q").

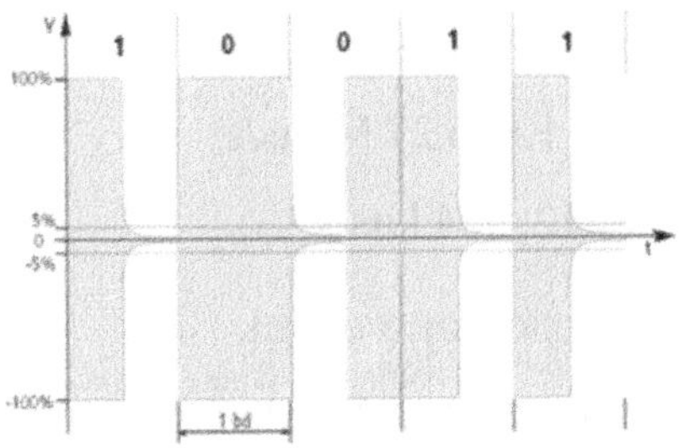

Figure 2.6. *100% ASK modulation*

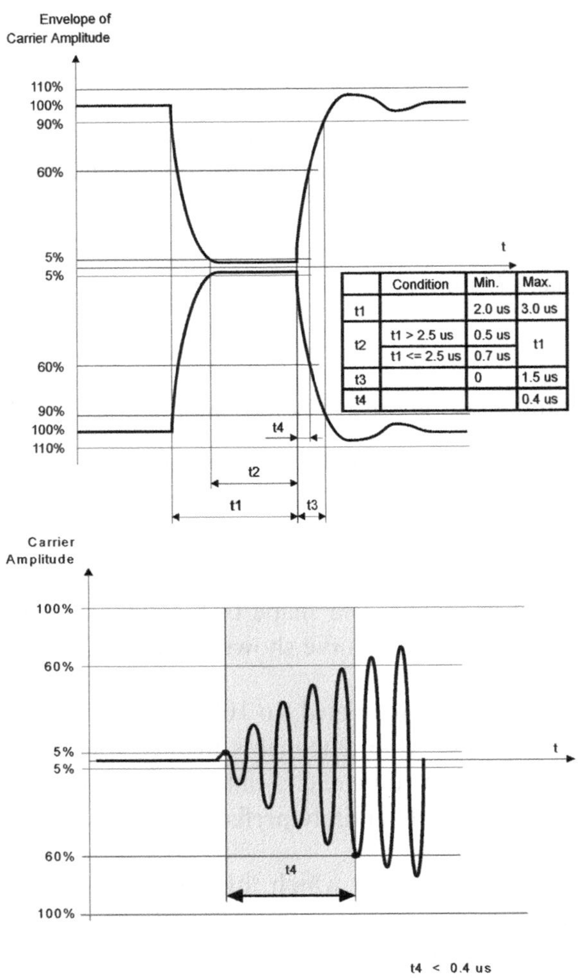

Figure 2.7. *Characteristics of starting and ending carrier modulation*

ASK modulation at 10%

Figure 2.8 indicates the theoretical rate of a carrier modulated at 10% ASK by an NRZ bit encoding such as ISO 14 443-2 type B.

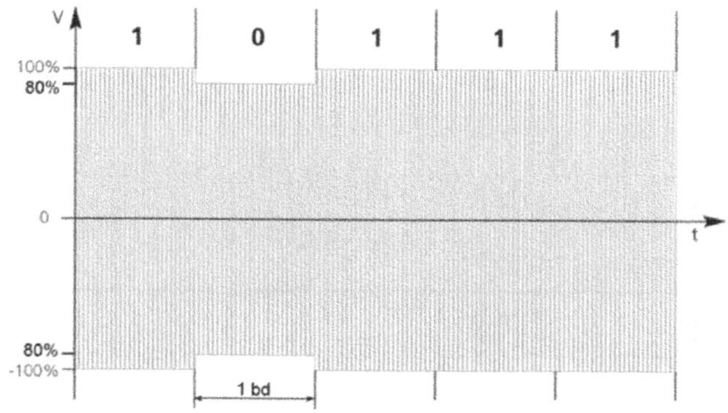

Figure 2.8. *Modulation at 10% ASK by an NRZ bit encoding*

Here too, since nothing is perfect, it is also necessary to define the shapes of the transitory signals that can be produced during the amplitude jumps.

Figure 2.9 also shows how to characterize these parameters for a carrier frequency of 13.56 MHz.

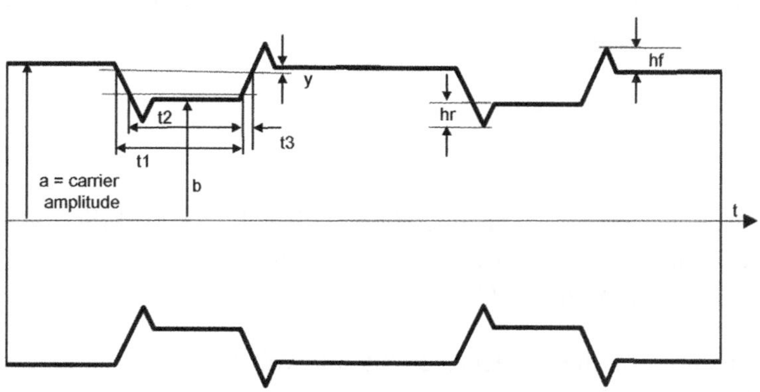

	Min (µs)	Max (µs)
t1	6,0	9,44
t2	3,0	t1
t3	0	4,5

y	0,05 (a-b)
hf, hr	0,1 (a-b) max

Figure 2.9. *Shapes of transient signals during amplitude jumps*

ASK modulation at 10%–30%

Figure 2.10 indicates the rate of a carrier modulated at ASK 10%–30% by a Manchester bit encoding such as that used in JIS X6319-4 – for FeliCa products.

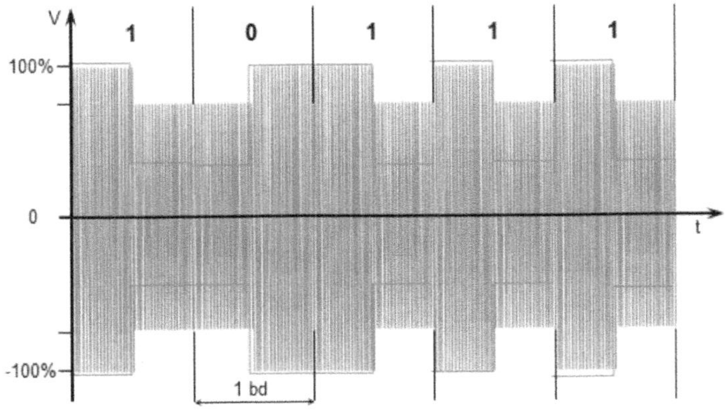

Figure 2.10. *Modulation at 10%–30% ASK by a Manchester bit encoding*

2.2.2.2. *Downlink*

During the downlink communication a target uses a load modulation (passive or active) enabling a bit to be created that does or does not generate a sub-carrier, as desired.

2.2.2.2.1. Bits encodings

Manchester coded sub-carrier (MSC)

The Manchester-coded bit encoding with sub-carrier–MSC – is a sub-coding variant of the Manchester encoding mentioned above (used in ISO 14 443-2 & 3 type A and NFC).

During most of the Manchester bit interval, the signal composing it is modulated by an all-or-nothing load modulation (OOK modulation) at an "fs" frequency called a sub-carrier. This fs frequency is generally an entire submultiple of the carrier frequency fc. Most of the bit's duration usually carries 4 or 8 periods of sub-carrier. As Figure 2.11 shows the bit encoding is therefore the result of the superposition of a Manchester coded signal and the presence of the sub-carrier.

EXAMPLE.–

fc = 13.56 MHz

fs = 423.75 kHz = fc/32

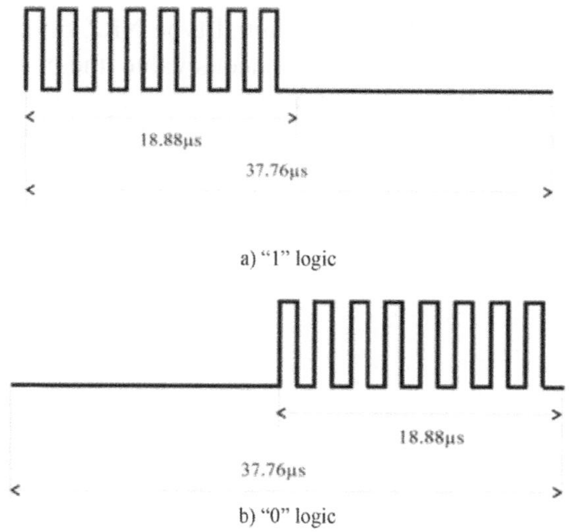

Figure 2.11. *Manchester coded sub-carrier – MSC*

Figure 2.12 shows an example of the MCS signal's OOK modulation action on the carrier supported by the initiator after the target's load modulation action.

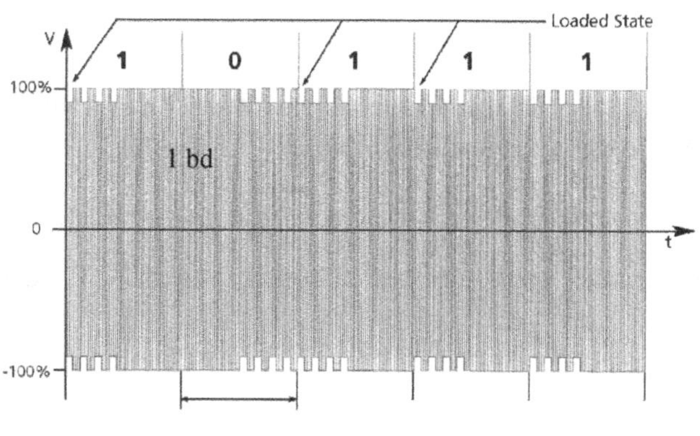

Figure 2.12. *MCS signal's OOK modulation action on the carrier*

The simultaneous RF reception of two bits of complementary value for this type of encoding creates an "ET logic HF" function, enabling third information known as "collision" to be defined.

Binary phase shift keying (BPSK)

The bit sub-coding BPSK uses an initial type NRZ bit coding, jumps by encoding, according to the bit's instantaneous value, the phase (0° and 180°) of a sub-carrier frequency generally results from the carrier frequency. This method has been retained for the transponder's downlink to the initiator in norm ISO 14 443–3 part B (Figure 2.13).

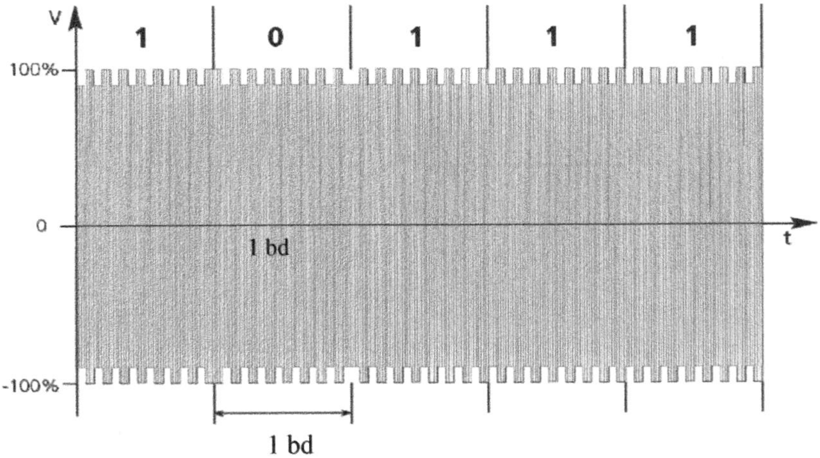

Figure 2.13. *Binary phase shift keying*

Manchester

The bit encoding for the Manchester return link is the same as that mentioned for the uplink (used in JIS X6319-4 – FeliCa)

2.2.2.3. *Summary of the consequences for antennae design*

Table 2.3 summarizes the normative constraints on NFC design.

2.3. Conclusion

We are really sorry to have inflicted many technical notions on some of you, which are long known and rehearsed with the systems for contactless smart cards that are an integral part of the bases and “legacies” borrowed from “the pre-existing” by the different NFC nuances (ISO, NFC Forum). These are however are strictly necessary for newcomers to this domain due to the newness of the media term “NFC”.

	Chip cards			NFC			Immediate consequences for antenna design
	ISO	EMV	ISO	ISO		NFC Forum	
	proximity		vicinity	IP1	IP2		
	14443	EMVCo L1	15693	18092	21481	Analog/DP	
14443-2	yes	yes		yes	yes	yes	
							Q for proximity
Bit rate in kbps		106 only		106 and 212-424	106 and 212-424	106 to 424	Q for vicinity
							Bp for 106
							Bp for above
Field H in A/m	1.5 to 7.5	not specified	0.5 to 5	1.5 to 7.5	1.5 to 7.5	not specified	EMV & Forum = not specified
Retro-modulation							ISO = EMV & Forum another possibility
14443-3	yes			yes	yes	yes	Target management, and thus stacks to be managed
Data collisions	yes	without anti-collision	yes	yes	yes	yes	Multi-card prohibited, and thus no stacks
RF collisions	no	No	no	yes	yes	yes	
Active mode	no	No	no	yes	yes	no	Dynamic initiator/target turnover
Tests	10373-6	EMVCo L1	10373-7			not available	
							ISO = field
Field H in A/m	yes	no	yes	yes	yes	no	EMV & Forum another possibility
Antenna class	yes (6)	1	not specified	not specified	not specified	yes (3)	
		3				3	
Volume	no	yes	no	no	no	yes	ISO = no volume imposed EMV & Forum two different volumes
Distance	no	yes	no	no	no	yes	ISO = no distance imposed EMV & Forum two different distances

Table 2.3. *Normative constraints on the design of NFC antennae*

3

Regulatory Constraints and Recommendations

3.1. Regulatory constraints specific to NFC and NFC antennas

A "regulation" consists of series of official documents/rules established by an organization attached to the "state" or a "community of states", which is compulsory to respect because of prescriptions, rules and regulations, laws, edicts and/or decrees, and other legal texts governing a social activity. The regulatory constraints to which NFC devices and the world associated with their antennas are subject are mainly environmental, and are generally present at four levels:

– radiation and RF pollution (ERC 70 03 and the norm ETSI 300 330);

– human exposure to EM fields;

– respect of private life and privacy – or *privacy impact assessment* (PIA)– (CEN EN 16 571);

– waste management (integrated circuits and antennas of the targets/tags).

3.1.1. *State of RF regulations*

All the NFC systems described in this book have antennas which can be placed at the level of the initiator and also at the level of the target. A great number of regulatory texts indicate, by way of allowable transmitted frequency values, authorized radiated levels, specific masks/templates, the constraints and restrictions (radiation, pollution, susceptibility, etc.) to which equipment devoted to contactless applications in the broadest sense (contactless chip cards, RFID, NFC, geolocation, etc.) are subject.

The problem regarding regulations of acceptable radiation levels and admissible levels of RF pollution is far from simple, owing to the great diversity of the laws and exceptions present in each region of the world and/or country. Put briefly, it is complex. A few global, American, European and national organizations govern and preside over the elaboration of parameters affecting short range devices (SRD) systems, which include NFC systems. These organizations include:

– at world level: the International Telecommunications Union (ITU) headquartered in Geneva in Switzerland;

– in the United States: the well-known US Federal Communications Commission (FCC) which, under the aegis of the American National Standards Institute (ANSI), published the famous reference document *US Code of Federal Regulations*, "Radio Frequency Devices", which is the local regulatory bible;

– in Europe, the recommendation in force is ERC – REC 70 03 "Relating to the use of Short Range Devices (SRD)" from the CEPT/ERO (*European Regulation Organization*), of which readers are strongly recommended to obtain the latest copy (from www.ero.dk, based in Denmark). In addition, the measuring and testing methods used by ETSI – "Electromagnetic compatibility and Radio spectrum Matters (ERM) – Short range devices": EN 300 – 330 – frequencies from 9 kHz to 25 MHz, and thus for 125 kHz and 13.56 MHz – follow the recommendations of the ERO;

– in Japan: the ARIB specifications are authoritative;

– in France: two organizations govern the usable frequency distributions and their uses;

- Agence Nationale des Fréquences (ANFR),

- Autorité de Régulation des Communications Electroniques et des Postes (ARCEP);

which refer as closely as possible to the ERC European Recommendations and produce documents serving as the basis for the elaboration of French norms and regulations pertaining to short range devices (SRD), non-specific, to which NFC applications are attached.

3.1.2. *Constraints pertaining to radiation and pollution by NFC*

In RF applications (RFID, NFC, etc.), one of the fundamental constraints is to respect the nationally- and internationally-agreed regulation levels of RF radiation and pollution. Despite the fact that at HF, it is possible to use various frequencies, today, practically everyone works with contactless chip cards and NFC at the same frequency (13.56 MHz). At that frequency, the regulations and norms on pollution are more or less in harmony throughout the world, and from region to region, there are a few minimal variations.

3.1.3. *The ERC 70 03 recommendation and the ETSI 300 330 norm*

3.1.3.1. *ERC 70 03*

Appendix 9 of ERC 70 03 is the official European regulatory document on "inductive applications" and the frequencies recommended for these applications (see Figure 3.1).

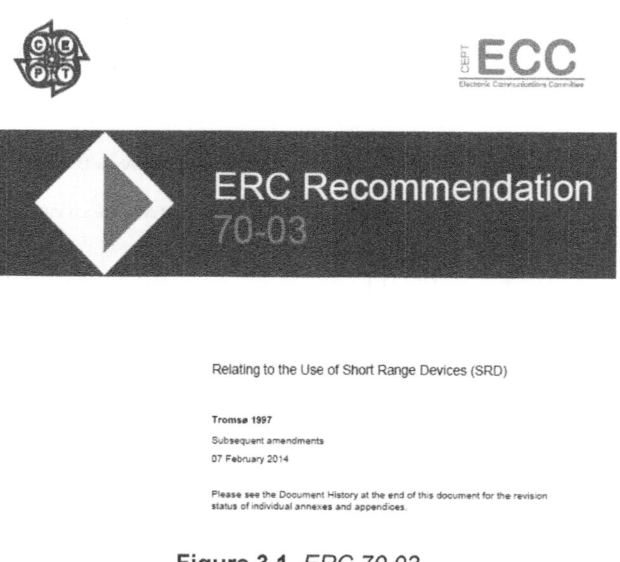

Figure 3.1. *ERC 70 03*

As usual, at LF and HF, the specification does not define the maximum allowable values of the radiated power, but of the radiated magnetic fields in dBμA/m (or electric fields in dBμV/m, depending on the country), measured at 10 m (or 30 m, depending on the country) in accordance with the recommendations of the ETSI 300 330 norm (see Figure 3.2).

ETSI EN 300 330-1 V1.4.1 (2004-11)

European Standard (Telecommunications series)

**Electromagnetic compatibility
and Radio spectrum Matters (ERM);
Short Range Devices (SRD);
Radio equipment in the frequency range
9 kHz to 25 MHz and inductive loop systems
in the frequency range 9 kHz to 30 MHz;
Part 1: Technical characteristics and test methods**

Figure 3.2. *ETSI 300 330*

3.1.3.2. *ETSI 300 330*

3.1.3.2.1. The template/spectrum mask

At the frequency of 13.56 MHz, the regulation requires that the spectral template given in Appendix L of ETSI 300 330 (see Figure 3.3) be respected.

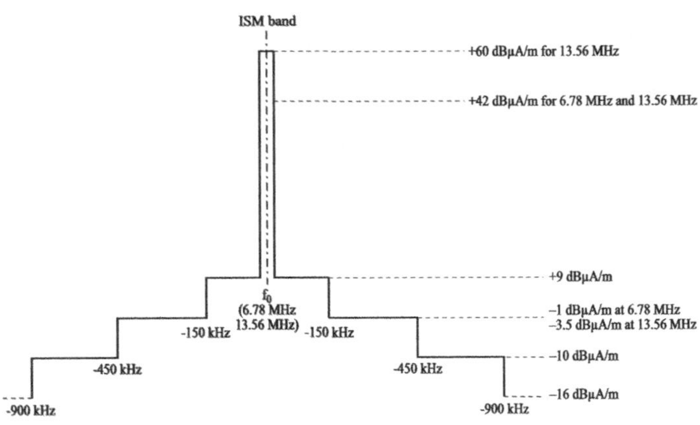

Figure 3.3. *Spectral template from Appendix L of ETSI 300 330*

3.1.3.2.2. Values

Beyond the functional normative aspect corresponding to the ISO 14 443 and NFC ISO 18 092 norms, as indicated above, it is also compulsory to satisfy the norms in force governing the levels (amplitude and template) of RF pollution, whose values are measured in Europe by the ETSI and regulated by ETSI 300 330, on the carrier frequency in far magnetic fields at 10 m, and in the United States by the FCC (FCC 47 part 15) in electric fields at 30 m; but whose value when expressed at 10m in terms of magnetic field is exactly the same value as that of the ETSI.

The values of the carrier frequency and of its maximum amplitude (in ETSI, around 42 dBμA/m at 10 m at 13.56 MHz +/- 7 kHz) are not too difficult to regulate. We have a quartz crystal in the application, and all is resolved.

The most difficult thing is to limit (reduction by around 60 dB) the levels of the high-ranking odd-numbered harmonics (produced by the origin of square-wave signals produced by the integrated circuit used), often considered to be "spurious", which fall into RF bands (generally of order N = 13 = 176.28 MHz at N = 27 = 366.12 MHz).

Table 3.1 indicates the maximum authorized values in Europe.

<table>
<tr><th>Frequency range (MHz)</th><th>H-field strength limit (H_t) dBμA/m at 10 m</th></tr>
<tr><td>6,765 ≤ F ≤6,95
13,553 ≤ F ≤ 13,567
26,957 ≤ F ≤ 27,283</td><td>42, note3</td></tr>
<tr><td>13,553 ≤ F ≤ 13,567</td><td>60 (see notes 2 and 3)</td></tr>
<tr><td colspan="2">Note 1: For the frequency ranges 9 to 70 kHz and 119 to 135 kHz, the following additional restrications apply ti limits above 42 dBμA/m:
- for loop coil antennas with an area > 0.16 m^2 table 3 applies directly;
- for loop coil antennas with an area between 0,05 m^2 and 0,16 m^2 table 3 applies with a correction factor. The limit is: table value + 10 × log (area/0.16 m^2);
- for loop coil antennas with an area < 0.05 m^2 the limits is 10 dB below table 3.
Note 2: For RFID and EAS applications only.
Note 3: Spectrum mask limits, see annex L.</td></tr>
</table>

Table 3.1. *Maximum authorized values in Europe*

EXAMPLE IN EUROPE.– In the case of a (true) power Pa (in watts) radiated on the carrier frequency by the initiator antenna (Pa = Ra I_ant^2, where Ra is the radiative resistance of the antenna), in "far fields", the value of the electric field E at "d" = 10 meters is equal:

$$\text{E\_ETSI\_10 m} = \frac{7 \times \sqrt{\text{Pa}}}{\text{d} = 10\text{ m}}$$

Let us translate this value obtained in V/m into dB μV/m, and then convert the result into the value of the magnetic field. Knowing that:

$$H = E / Z_0 \text{ in A/m where } Z_0 = 377 \text{ ohms} = \text{no-load impedance}$$

we can write, in dB:

$$20 \log H = 20 \log (E/Z0)$$

$$H \text{ (in dBA/m)} = E \text{ (in dBV/m)} - 20 \log Z0$$

$$H \text{ (dBA/m)} = E \text{ (in dBV/m)} - 20 \log (377)$$

This gives us the well-known equation, in far field and in the equatorial plane:

$$H \text{ (dBA/m)} = E \text{ (in dBV/m)} - 51.5 \text{ dB}$$

Very often, in view of the absolute values of the radiated fields, we use their equivalent values expressed in dBμA/m for *H* and in dBμV/m for *E*, so:

$$H \text{ in dB}\mu\text{A.m}^{-1} = 20 \log \frac{H \text{ in A} \cdot \text{m}^{-1}}{1 \times 10^{-6}}$$

so, again:

$$\text{H dB } \mu\text{A/m} = (\text{E in dB}\mu\text{V/m}) - 51.5 \text{ dB}$$

Furthermore, in order to adhere to ETSI 300 330, we must not surpass the maximum value of the field H, measured at 10 m:

$$\text{H\_ETSI\_10 m (in dB}\mu\text{A/m)} = \text{around } 23\ \mu\text{A/m}$$

We know (see Chapter 1) that H_10 m = H_0 / [(1 + (d/r)²)^3/2], so it is possible to work back to H_0.

To remain within the limits of ETSI 300 330, we can write:

$$\text{E\_ETSI\_10 m (in dB}\mu\text{V/m)} = \text{H dB } \mu\text{A/m} + 51.5 = 74.5 \text{ dB}\mu\text{V/m}$$

or indeed:

E_ETSI_10 m = 5.31 mV/m

The constraints do not stop at satisfying the maximum authorized values (of radiated fields E or H or power ERP or EIRP, or indeed power spectral density depending on the norms) but also bring the whole of the radiated spectrum into the template (or the *mask*) imposed by ETSI 300 330, and maintain the radiation level of the lateral bands of the signal, which is one of the crucial points determining the satisfaction (or not) of these norms; by means of the bandwidth, and thus the value of the quality coefficient Q of the antennas and the choice of EMC filtering techniques. However, the form and content of the emitted spectrum depends mainly on the type of bit coding, the data rate ensuring communication, the type of modulation, the principle of processing and collision management, "randomization", and recurrence of the signal; but also on the possible distortion of the signal if it is deformed by improperly-designed EMC filters. Additionally, see Figure 3.4 which will shed a certain amount of light on the matter.

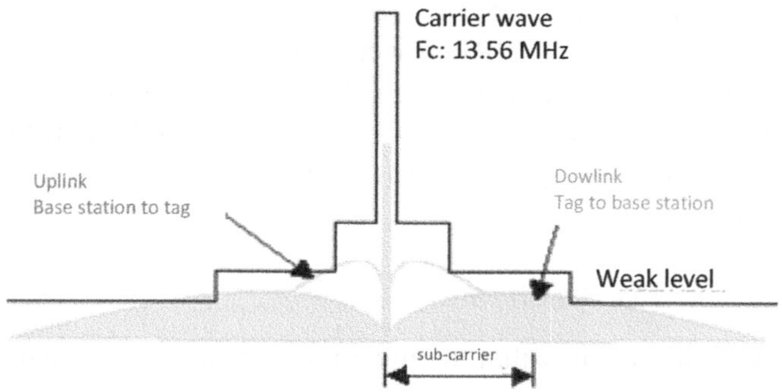

Figure 3.4. *Respective shapes/values of the spectra for uplink and downlink*

3.2. Constraints due to recommendations

A few more things!

Certain entities, associations, etc., do not have the weight to directly impose certain things on a state but, on the other hand, they do have a great deal of competency to recommend certain values and criteria in their domains of expertise. States are free to accept, recommend, adopt, impose, etc., these values by laws or decrees. The example below will illustrate all of this.

3.2.1. *Exposure of the human body to EM fields*

In view of health constraints, it is also necessary to limit human exposure to maximum values of EM fields.

The organization in charge of these problems is the International Commission on Non-Ionizing Radiation Protection (ICNIRP). Also note that ICNIRP regularly meets with the World Health Organization (WHO).

Thus, the ICNIRP has outlined "recommendations" which must not be exceeded (e.g. for a period of 6 minutes) for the head and torso, using a specific absorption rate (SAR) of over 2 W/kg for 10 g of human tissue with signals of up to a strength of 300 GHz. In the United States, the ANSI and IEEE documents specify 1.6 W/kg for 1 g. Generally, with contactless technology, it is "long-distance" systems that are most critical in terms of respecting these recommendations, because they require the emission of higher levels of ERP and/or EIRP power. It should be noted that the content of these documents was taken into consideration in the drafting of norms for proximity (ISO 14 443), vicinity (ISO 15 693) and thus NFC ISO 18 092, so systems which satisfy these norms also conform to the recommendations on human exposure.

3.2.1.1. *Specific absorption rate (SAR)*

It is not difficult to show that the most representative parameter, the SAR, is equal to:

$$\mathrm{SAR(rms)} = \frac{\sigma\left|E(\mathrm{eff})\right|^2}{\rho} = \frac{J^2}{\rho\sigma} = \frac{(\sigma\omega RB/2)^2}{\rho\sigma}$$

where:

– ρ = density of human tissue in g/cm^3 = 1000 kg/m^3;

– σ = conductivity of human tissue in S/m = 0.51 S/m;

– ω = pulsation in rad/s;

– B = magnetic induction in Tesla;

– R = constant 0.2.

To situate a value of the SAR, here is an example, taken for a proximity application (ISO 14 443 and/or NFC ISO 18 092).

Supposing that, at a distance of 20 cm, we (still) have a magnetic field H produced by the initiator of 1.33 A/m:

$$SAR = \frac{\left(0.51\times 2\pi\times 13.56\times 10^{6}\times 0.2\times 4\pi\times 10^{-7}\times 1.33\times 0.5\right)^{2}}{1000\times 0.51} = 0.103\ \text{W/kg}$$

This value of the SAR concerning whole-body exposure in the work environment thus conforms to the recommendations pertaining to human exposure (for information, the SAR value of a mobile telephone is between 0.7 and 1 W/kg).

For example, for a power of around 0.6 W, which gives an effective radiated power of around 20 μW, we can construct Table 3.2.

Measuring distances	SAR in for 1 g	SAR for 10 g
mm	W/kg	W/kg
5	around 0.16	around 0.1
15	around 0.08	around 0.05

Table 3.2. *Example of SAR for an effective radiated power of around 20 μW*

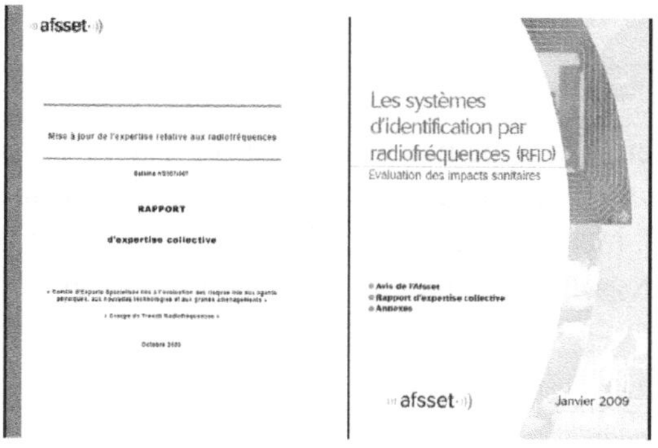

Figure 3.5. *Examples of ANSES documents (formerly AFSSET)*

Given the low power values involved by initiators in the case of NFC at 13.56 MHz, the problem of human exposure to non-ionizing radiation does not arise. For other applications, initially, readers are recommended to consult the two excellent documents available on the Website of ANSES in France (formerly AFSSET) – see Figure 3.5.

3.2.2. *Societal constraints due to individual freedoms (privacy)*

Whether RF systems work with near fields or far fields, endless debates can be conducted about individual freedoms and the respect of privacy.

3.2.2.1. *Mandate 436*

In 2008, the European Commission, based in Brussels, published "Mandate 436" (see Figure 3.6), dealing with all the details of the problems raised by the problems of privacy, individual freedoms and societal aspects of RFID, of which NFC is only one specific sub-branch.

EUROPEAN COMMISSION
ENTERPRISE AND INDUSTRY DIRECTORATE-GENERAL

Innovation policy
ICT for Competitiveness and Innovation

Brussels, 8 December 2008
DG ENTR/D4

M 436 – EN

STANDARDISATION MANDATE
TO THE EUROPEAN STANDARDISATION ORGANISATIONS CEN, CENELEC AND ETSI
IN THE FIELD OF INFORMATION AND COMMUNICATION TECHNOLOGIES
APPLIED TO RADIO FREQUENCY IDENTIFICATION (RFID) AND SYSTEMS

Figure 3.6. *Mandate 436*

Sooner or later, it will be necessary and obligatory to establish a *privacy impact assessment* (PIA) in accordance with the European Norm EN 16571 (from mid-2013), meaning detailing all the risks linked to the *distribution of impacts on privacy* of a system, and send it to the local CNIL[1] for approval.

---

1 The *Commission nationale de l'informatique et des libertés* is in charge of ensuring that computer technology serves citizens' needs and that it does not impinge human identity, human rights, privacy, individual freedoms, or public liberties. Similar regulating bodies exist in other countries.

Thus, beware of quirky application and even more quirky antennas in NFC to satisfy those applications.

3.2.2.2. *Privacy impact assessment: "EN 16 571"*

Below, for information and for immediate obligatory applications, are a few extracts from the "scope" of the PIA:

"On 12 May 2009, the EC published a recommendation (mandate M436) on the implementation of the principles of respect of privacy and data protection in applications based on RFID (regardless of the application).

By virtue of this recommendation, a framework for evaluating the impact of RFID applications on personal data and privacy, developed by businesses, must be submitted for approval to the "Article 29" Working Group on data protection. This evaluation is commonly called a privacy impact assessment (PIA). The framework for assessing the impact of RFID applications on privacy constitutes a response to this requirement.

The PIA process is based on risk management in terms of the respect of privacy and data protection, which essentially hinges on the implementation of the EU's RFID recommendation, and conforms to the best practices and the EU's legal framework. It is designed to help the operators of RFID applications, in the broad sense, detect the risks to privacy associated with an RFID application, to assess their probability and document the measures to be taken to deal with them. Such impacts (if there are any) may vary significantly depending on whether or not the RFID application handles personal data. The PIA framework gives RFID operators guidelines regarding the methods of risk assessment; in particular, it offers appropriate measures to limit any probable impact on privacy or data protection effectively, concretely and proportionately.

Finally, the PIA framework is sufficiently general to cover all RFID applications (including NFC), whilst dealing with the peculiarities and specific points at sector- or application level.

The PIA framework fits into the context of other norms guaranteeing information, data management and exploitation, which encourage good data governance for RFID and other applications. The current framework could serve as the basis for the development of PIA models dedicated specifically to an industry, a sector and/or a particular application".

With this now clarified, it is time to move on!

European norm CEN EN 16 571 gives a detailed description of the content, procedure and methodology to satisfy a PIA. In Europe, the CNRFID – *Centre national pour la RFID* – a French center that has made significant contributions in this area, has been delegated by Brussels to help fill it in using specific software tools.

3.2.2.3. *Privacy by design*

Obviously, the ultimate goal is not to reach the end of one's project and then think about a PIA, but instead to bear it in mind throughout the whole process of design and development of an NFC device. This approach is known as *privacy by design*. Next, the chain of privacy protection must be examined, checked, verifiable and quantified, and guaranteed from end to end.

3.2.3. *Environmental constraints*

Whenever we speak of RFID, contactless and NFC, the question of the recycling of the (potentially very numerous) tags and targets eventually comes into the debate. The silicon chip in the integrated circuit is small, and we know how to process and recycle them, but what should be done with the antennas of NFC initiators and targets which, in the long term, will become an ever-worsening source of waste?

3.2.3.1. *Waste management as it relates to the waste electronics of NFC antennas (from initiators and targets)*

Electrical and electronic equipment (EEE) often contains substances or components which are dangerous for the environment, and have great potential for the recycling of the materials from which they are made (ferrous- and non-ferrous metals, rare metals, plastics, etc.). For example, in France, the Ministry of Ecology, Sustainable Development and Energy is in charge of regulating waste electrical and electronic equipment (WEEE). These environmental issues have led to the establishment of an industry for the specific management (collection and recycling) of such WEEE, based on the principle of continued responsibility of the manufacturers of this equipment.

3.2.3.1.1. Regulation and organization of this industry

The directives on waste electrical and electronic equipment and dangerous substances contained in that equipment define the conditions of marketing of EEE, and the framework for WEEE. In summary, they set the following main principles:

– distinction between household equipment and professional equipment;

– prohibition from using six dangerous substances in EEE;

– selective collection of WEEE;

– selective systematic treatment of dangerous substances and components contained in EEE;

– achievement of objectives of reuse/recycling and value creation of WEEE;

– extended responsibility on part of the manufacturers for the management of WEEE stemming from the equipment which they produce;

– means of treatment of WEEE relative to the composition of the EEE in question and elimination of waste issuing from that equipment;

– conditions of implementation of the obligations of collection of used EEE by the distributors;

– conditions that a mandate holder must fulfill under the terms of the environmental code in order to ensure the respect of the obligations incumbent on the producer who granted that mandate.

3.2.3.1.2. Labeling of electrical and electronic equipment

European norm EN 50419 stipulates labeling for EEE which applies as long as the equipment in question is not a subsidiary part of a different type of equipment. The norm gives an indicative list of the products in the various categories, which serves to clearly identify the producer of the device and certify that it has gone on the market after August 2005. The definition of a technical means to identify the manufacturer, such as a barcode, an electronic data support or a microchip, is not covered by this norm.

3.3. Constraints of the NFC market

3.3.1. *NFC applications of short range devices*

The reason we create a product is that we hope to sell it. Hence, it needs to be “sellable” (and better still, “buyable”!), and its function, its usefulness, its sale price on entry into the market, the lowering of that price over time, its profitability, etc., need to be known. In short, it is important to study the “business model” in great detail before starting out.

Let us briefly recap that the market for *short range devices* (SRD), devoted to “contactless” RF applications, of which NFC is a part, is broadly subdivided into two major areas:

– the markets for “niche” applications, where products are specifically developed for one or more types of applications of small or medium production, where we tend not to worry too much about costs and prices, and generally, sell added value;

– the markets for “mass” applications which, as the name indicates, are high-volume markets (specific or non-specific products), where the constraints of costs and prices are of prime importance, because it tends to be a cheap price or an attractive discount, rather than the products themselves, which draw in buyers.

Generally, we choose the field we wish to work in, inject the necessary means of development, production, etc., as required by the target market, and deal with the economic constraints inherent in that market.

3.3.2. *Costs and market prices desired by users*

Often, NFC systems are intended for the consumer-electronics market, and thus for widespread distribution/mass market, so cost is crucial. Here, then, is a brief description of the technico-economic chain established to satisfy users’ expectations and desires:

inexpensive → batteryless → remotely powered → radiated power authorized

– inexpensive, so without a battery on board the element and no problem with connectors, along with devices to guard against vibrations and bad contacts;

– the term “batteryless” here means that the local power supply to the system must be performed by a principle known as “remote power” with a near HF field radiated by the initiator;

– one of the questions we may legitimately ask is whether it is possible to remotely power the target whilst still respecting the constraints of the RF regulations in force (bands, frequencies, powers and/or maximum authorized radiated fields H or E, template (thus data rates, bit coding, collision management, usable duty cycle, etc.).

Once we have read and studied all these documents, and understood all of the ins and outs of the texts and the inevitably small print, it is necessary to focus on the technology of the semi-conductors to discover the results, in terms of operative distance, of marrying the available technologies, regulations, and remote-power procedure.

3.3.3. *Beware of false advertising*

To conclude this chapter, and to call a spade a spade, before proudly declaring a product to be an NFC device and adding “conforms to NFC ISO 18 092 or 21 481 norm or the NFC Forum standard” (of course, in active or passive mode, batteryless or battery-assisted, etc.), in the judicial and legal sense (otherwise, we leave ourselves open to prosecution and other disciplinary procedures for misleading advertising), we need to be certain that it is at least worthy of that certification, and that it has passed numerous conformity tests, rather than simply being a simple ISO 14 443 system (though there is nothing pejorative about this), as often happens through incorrect use of language.

PART 2

Constraints Due to the Field of Applications of NFC

Introduction to Part 2

In order to discuss the technical constraints associated with different applications, it is first of all necessary to explore them and describe them succinctly! That is the aim of this second part of the book, during which we will analyze the consequences and the applicative constraints of the technical principles that support the technique, technology, typologies and the application fields of the NFC. The following two chapters are as follows:

– Chapter 4: “Applicational Typologies of the NFC and their Consequences”;

– Chapter 5: “Constraints due to Fields of Application”.

4

Applicational Typologies of the NFC and their Consequences

We will start with a description of the typologies usable with the NFC concepts, then describe the constraints due to the application field stemming from these typologies.

4.1. Applicational typologies of the NFC

No, no and no – the applicational typology of the NFC is not linked only to mobile telephony! No, the NFC is not merely a contactless payment card (emulator). No, because these views are highly reductive, and there are numerous applicational typologies of the NFC, which we will now briefly review, along with the consequences for the design of the NFC device itself. Once gain, in order to describe these typologies we will have to go back over several keywords of vocabulary specific to the NFC from the definitions of the norm ISO 18 092 (communication protocols NFC IP1 and IP2 in the spirit of the interface), of the NFC Forum specifications.

4.1.1. *Technologies*

Beyond the ISO NFC IP2 and IP1 norms, the NFC Forum defined three generic groups of transmission types under the term "technologies".

4.1.1.1. *NFC-A technology*

"NFC-A" technology is very similar to ISO 18092 NFC IP 1 and IP 2, which were heavily inspired by the contactless chip card norm ISO 14443 type A, parts 2 and 3. In the NFC Forum, NFC-A is used for the tags called T1T, T2T, T4AT and P2P.

4.1.1.2. *NFC-B technology*

"NFC-B" technology is very similar to ISO 14443 type B, tirets 2-3, which had already been reused solely in ISO NFC IP 2. In the NFC Forum, NFC-B is used for the tags called T4BT and P2P.

4.1.1.3. *NFC-F technology*

"NFC-F" technology is very similar to ISO 18092 NFC IP 1 and IP 2 which, in large part, were heavily inspired by the Japanese norm JIS X6319-4. In the NFC Forum, NFC-F is used for the tags called T3T and P2P.

4.1.2. *"NFC Forum Devices" and "NFC Forum Tags"*

Even more new terminology.

4.1.2.1. *NFC device*

The ISO norms pertaining to NFC use this term to denote a communicative "initiator" or "target", be it in active mode or in passive mode (see section 4.3). For its part, after dividing up different classes of "technologies" (see section 4.1), the NFC Forum also decided to define and separate all "NFC devices" (from the ISO) into two classes of elements:

– *NFC Forum Devices*;

– *NFC Forum Tags*.

4.1.2.2. *NFC Forum Devices*

Whatever their physical forms (a mobile phone, PC, TV, camera, photo frame, tablets, etc.), *NFC Forum Devices* are elements which support the different modes and roles of operation described below, for all products and norms supported by the NFC Forum, which implement at least the *mandatory* parts of the protocol *stack* issued by the NFC Forum, and satisfy the demands and necessities of interoperability imposed by the NFC Forum.

NFC Forum Devices must:

– (*mandatorily*) communicate with one another in accordance with ISO 18092, the protocols attached to the NFC Forum LLCP, "P2P", read (and write) the different types of NFC Forum Tag and be able to function as an initiator, target and reader/writer;

– should (*optionally*) offer the possibility of emulating a contactless card. Thus, an NFC Forum Device can be used as a contactless chip card of type ISO 14443 A (such as MiFare) or B or FeliCa (via the platforms T3T, T4AT or T4BT).

As we can see, an NFC Forum Device is a technologically complex element (and so is its antenna), because it is mandatory for it to:

– implement multiple technologies (A, B and F) on board;

– be able to switch from listen mode to poll mode;

– be able to discover whether there are other NFC Forum Devices or NFC Forum Tags in its vicinity, and process RF collisions and data collisions;

– communicate with those other devices.

4.1.2.3. *NFC Forum Tags*

Once again, no form factor is specified for these elements, but often a case of use is envisaged: that of a *smart poster*, which essentially means nothing and tells us nothing about its form factor – is it round, square, small, large? It is a mystery. More formally, an NFC Forum Tag is a contactless tag or a (smart) card supporting the *NFC Data Exchange Format* (NDEF) protocol on a passive communication, and consequently, an NFC Forum Tag only transmits responses and supports at least one legacy communication protocol, defined in the specifications of the NFC Forum.

Let us now briefly look at the definitions of the "modes" and "roles" which NFC devices may take on during their respective operations.

4.1.3. *"Modes" of communication of an NFC Forum Device*

By dividing its time, an NFC Forum Device can support different "modes" of operation. There are four such modes which pertain to communication between elements. However, the word "mode" also covers two different, unrelated concepts.

To begin with, two sub-options of *modes of communication* are described in the norms NFC IP-1 (ISO 18092) and IP2 (ISO 21481).

Furthermore, two other modes – *listen mode* and *poll mode* – define how communication is established.

4.1.3.1. *Passive communication mode*

In this mode, the NFC Device "initiator" generates an RF field, which it modulates in order to send commands. The NFC devices' "*targets*" present in the field send their answers by load modulation (see Figure 4.1).

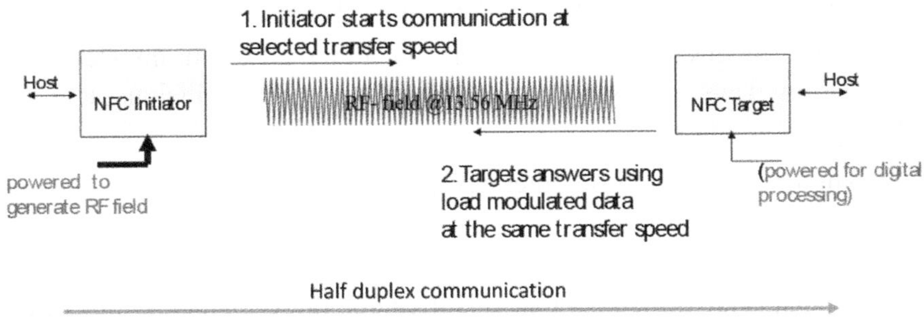

Figure 4.1. *Passive communication mode*

4.1.3.2. *Active communication mode*

In this mode, the NFC devices take it in turns to generate an RF field. The initiator begins by generating a field to send commands, and then stops completely. Then, in turn, the target produces its own field and modulates to send its answers (see Figure 4.2).

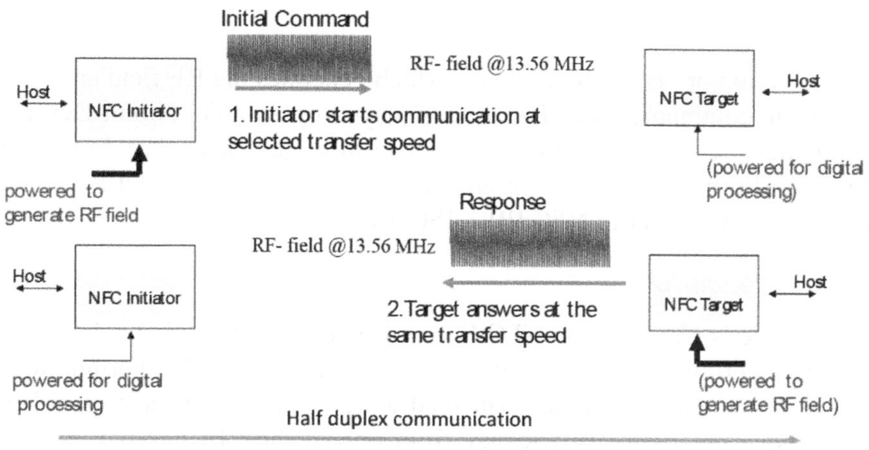

Figure 4.2. *Active communication mode*

NOTE.– These communication modes are completely independent of how the NFC devices are powered. There is no relation with the notions of remote power supply (batteryless) or battery-assisted operation of the NFC devices.

NOTE.– Whether the NFC devices (the initiator or the target) are in active or passive mode, transport of the data in the two directions takes place in an alternating manner, rather than simultaneously. Thus, communication is *always half-duplex.* This remark remains applicable for the whole of this book.

NOTE.– Although many people use the term *peer-to-peer*, abbreviated as P2P, to speak of "active mode" the term *appears nowhere in the ISO norms.* Indeed, "active mode" does not, in itself, constitute a true P2P model. A P2P model is a *client–server*-type model, in which each client is also a server. This is not exactly the case, because in active mode, there is always the same server: the initiator which started off the exchange remains in charge of that exchange for the whole duration of the communication, and the client remains a target.

4.1.4. *Role of an NFC Forum Device*

The NFC Forum defined four different roles for NFC Forum Devices, depending on the mode and communication protocol used.

4.1.4.1. *Initiator*

The "initiator" is the NFC device which generates the RF field and which begins the communication. This role is played by an NFC Forum Device in poll mode, having passed through a number of stages (*activities*). In this role, the NFC Forum Device communicates by using the NFC-DEP Protocol (which is very similar to NFC IP1 – ISO 18092).

4.1.4.2. *Reader/writer*

This role is played by an NFC Forum Device in poll mode having passed through a certain number of activities. In this role, the NFC Forum Device acts as a "contactless reader" (in read/write mode), using a protocol and commands "inherited" as a legacy from one of the "Technology Subsets" (platform T1T, T2T, T3T or T4T).

4.1.4.3. *Target*

An "NFC device" becomes a target when it responds to the initiator's commands, replying by way either of load modulation (with the RF field always being generated by the initiator) or modulating its own generated field. This is a role played by an NFC Forum Device having passed through a number of activities. In this role, the NFC Forum Device communicates by using the NFC-DEP Protocol.

4.1.4.4. *Card emulator*

The role of a card emulator is achieved when an NFC Forum Device is in listen mode. The NFC Forum Device then behaves like one of the technology subsets and a legacy protocol (T3T or T4T platform). In this role, the NFC Forum Device communicates using the ISO-DEP Protocol.

4.1.5. *Applicational typological possibilities and their constraints*

The large range of "active" and "passive" communication modes that support the NFC concept, alongside the numerous *batteryless* and *battery-assisted* power supply variations, make it possible to envisage a large number of applications of the NFC principle in numerous markets for numerous people with very specific economic interests and uses; thus with a large number of architectural problems and numerous technical, security, etc. constraints to solve – and we are only just beginning the NFC story!

4.1.5.1. *Applications additional to and/or moving away from the NFC norm*

Among the applications not officially described by the terms used in the ISO norms, but visible in the NFC Forum standard, is a widely available applicational field based on the possibilities of passive communication and a power supply that can be either *batteryless* or *battery-assisted.* This application consists of making an NFC system work in a "contactless chip card reader mode" with the associated concerns a little later.

Tables 4.1 and 4.2 summarize the range of "reader" applications (RFID, P2P, battery-assisted passive card emulation, remote-powered passive card emulation, etc.) engendered by the physical fundaments of the NFC system.

	Initiator *Batteryless*	Initiator *Battery-assisted*	
		Passive communication	Active communication
Target *Batteryless*		Initiator = Reader Target →Card reader Target →Tag	Peer-to-peer Pairing
Target *Battery-assisted*		Emulation card + crypto	Peer-to-peer Pairing explore

Table 4.1. *Range of application examples*

4.2. Application consequences and their direct constraints

From these generic possibilities are derive applicational services, which are generally grouped into the following for main marketing terms: "*Touch & Go*", "*Touch & Confirm*", "*Touch & Connect*" and "*Touch & Explore*" (Figure 4.3).

	NFC device									
		Initiator	Target							
Means of communication		Role		Role						
		Initiator		Tag	Card		Down-link			
Passive Half-duplex Bidirectional	P	Reader/writer	L	1T, T2T, T3T or T4T poster			PLM *ALM*		Batteryless	Poster, label
	P	Reader/writer	L		T3T or T4T chip		PLM *ALM*		Batteryless	Ordinary chip card. Bank card, loyalty card
		Reader/writer	L			T3T or T4T card emul	PLM *ALM*		Batteryless	Phone with flat battery, loading effect
						T3T or T4T card emul	PLM *ALM*		Battery assisted	Normal phone, loading effect
							ALM not described in NFC transmits		Battery assisted	
Active Half-duplex Bidirectional	P	Initiator NFC-DEP Protocol NFC IP1 - ISO 18 092			Other device				Battery assisted	

Table 4.2. *Range of NFC applications*

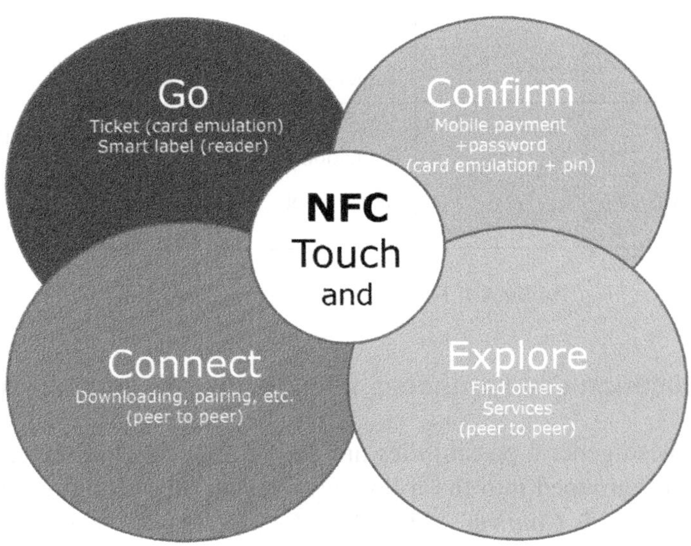

Figure 4.3. *"Touch & Go", "Touch & Confirm", "Touch & Connect", "Touch & Explore"*

4.2.1. *Touch & Go typology*

In "Touch & Go" applications and services, the user presents to the initiator a target (NFC target device in the broad sense – ticket, token, badge, card, mobile phone) able to trigger an action without having to confirm validation, for example for:

– a simple access control;

– verification of a ticket (for transport or entry to an event, for example);

– on the fly input of a URL contained in a "smart" label, itself included in a poster or advert;

– a ticket office where the user simply places the device containing the electronic ticket or access code close to the reader, etc.

4.2.2. *Touch & Confirm typology*

"Touch & Confirm" applications require an NFC device containing a human-machine interface (for example buttons, a screen, keyboard, etc.) to be presented to the initiator, through the interface of which the user can perform any action or gesture required to confirm the interaction by entering a code or a password, or simply agreeing to or validating the transaction. Applications characteristic of this mode of operation are for example:

– mobile payment using an NFC device (mobile phone, etc.);

– secure access control.

4.2.3. *Touch & Connect typology*

In the applicational field of "Touch & Connect" two NFC devices are placed face to face so that they can communicate with one another and transfer data in a "peer-to-peer" manner, for example:

– remote upload (or download) of data;

– exchange of music or images (for example from a mobile phone to a photo frame or a television screen);

– exchanging plane tickets from mobile to mobile or to a fixed terminal;

– intuitive synchronization of address files between two NFC systems (telephone to a PC), etc.

4.2.4. *Touch & Explore typology*

In the field of applications for "Touch and Explore" the NFC devices can propose several functions to the user. The consumer will thus be able to explore the capacities of the device to discover its functions and proposed services.

This concludes the brief overview of different applicational typologies of the NFC and their (vast) fields of application. Let us now move on to the range of different applications of the NFC though typology and their consequences for the design of NFC devices.

5

Constraints Due to Fields of Application

In the previous chapter, we defined the main classes of typologies. These cover numerous fields of NFC applications, which we shall examine now.

5.1. Range of technical possibilities of applications

In spite of what one may think, mobile telephony is not the only thing involved on the NFC market. NFC has a wide variety of applications, and it is very difficult to establish the "ecosystem" associated with NFC, because there is so much interlinking, inter-applications and overlap, both at a financial level and in terms of technical solutions to be implemented and problems to be resolved.

Now we shall examine the generic range of technical possibilities and application constraints for NFC, in terms of architectures, electronic functions, physical shapes, and associated form factors, which result from a whole host of applications for NFC devices – both initiators and targets. Though the following is by no means intended to be an exhaustive list, here are a few examples.

5.1.1. *In architectures and electronic functions*

5.1.1.1. *Interrogator*

The interrogator function is that of a device (in initiator mode) which, for example, can read simple tags and find their content, but cannot modify that content – e.g. a poster or billboard on the street.

5.1.1.2. *Reader*

The term “reader function” is fallacious, because it often includes the double meaning of “reading *and* writing”, rather than just reading. This being the case, the purpose of the NFC device is to read content, but also to be able to modify that content – for instance, in the supply chain.

5.1.1.3. *Target*

Caution must also be exercised with the word “target”, because in terms of functions, a target may be very powerful – e.g. a target in active mode (i.e. which is capable of transmission) or a target in passive mode (a simple tag), or indeed in card-emulation mode.

5.1.2. *Shapes, dimensions and form factors (publicity aside)*

Table 5.1, in two dimensions gives numerous examples of the technical diversity of NFC applications, and of the antenna shapes which are now beginning to be used. Readers could also add a third dimension to this table by including the respective typologies for these applications.

5.1.3. *Remarks and some clarifications about the table*

Some of the below applications classified in the general family of NFC are very commonplace, but are worth looking at in greater detail.

Owing to the great proximity and similarities between the various norms for “contactless chip cards” (in the ISO’s SC 17 series of norms), “RFID” (SC 31 series), and “NFC” (SC 06 series), the applications to which we refer here, are those which employ the function of simple tags, of type NFC IP1 (ISO 18092) and/or IP2 (ISO 21481) supporting the vicinity card norm (ISO 15693). With these applications, the devices work for a certain period of time either in pure RFID mode to read data at a medium distance (30 cm) or a long distance (70-80 cm), or for short distance (only a few cm), using a NFC mobile phone, either in reader mode or in card-emulation mode for certain tasks. In the interest of absolute clarity, let us discuss two examples which appear in the table.

		Architectures			Examples (non-exhaustive list)	
		Initiator	Target			
Markets	Functions	Writer/ reader	Tag	Chip-card emulation	Antenna dimensions (approximate)	Non-exhaustive examples of manufacturers
Loyalty cards	Chip cards, badges		✓ ✓		5 × 6 cm	
Mobile telephony	Smart phones, all applications	✓		✓	6 × 6 cm à 3 × 3 cm	Google, Apple, Samsung, etc.
	Add-on cards			✓	2.5 × 2.5 cm	Mercury
Bank payment	Smart card			✓	5 × 7 cm 2.5 × 7 cm	Gemalto, Morpho, Oberthur, G&D, etc.
	MPOS – Mobile Point of Sales	✓			5 × 4 cm screen size	Ingenico, Wincor Nixdorf, etc.
	Tablets, wallets	✓	✓	✓	above and below	
	ATM	✓			Φ = 7 cm EMVCo	Thalès, etc.
Industrial	Badges		✓		absolutely any shape and size	
	Labels		✓			
	Smart poster		✓			
	Number signs for races		✓			Dag System, etc.
Medical	Identification, tracking	✓	✓		4 × 1 cm	Trixell, Maquet
	Decontamination tank	✓			30 × 30 cm	IMS
	Monitoring of baby mice, ants, bees	✓	✓		very small 4 × 1 mm	Lutronic, Arelis, Nonatec, etc,
Consumer	Photo frames		✓		30 × 20 cm Screen size	Parrot, etc.
	Speaker systems		✓			Parrot, etc.
	Television sets	✓		✓	Secure access	Philips, etc.
	PC	✓		✓		Dell, etc.
	Advertising items	✓	✓		All formats	Strapmedia
	Museum guide		✓		Φ = 2.5 cm	Strapmedia
	Wireless charger for library	✓			10 × 8 cm	Samsung, Nokia Tagsys, etc.
Computing	USB dongles	✓	✓	✓	1 × 1.5 cm	Mercury, Neowave
	USB sticks		✓		2.5 × 2.5 cm	Neowave
	Terminals	✓				Taztag
Automobile	Car dashboards					Herman, Delphi, Valéo
	Ignition keys		✓	✓	3 × 5 cm	Conti, Valéo
	NFC battery charger	✓			10 × 8 cm	Delphi, Conti
Public transport	Ticket validation				Φ = 10 cm	Xeros, Parkeon Thalès, etc.
	Monitoring device				5 × 3 cm	
IoT – connected items	Watches, keys				2.5 × 2.5 cm Φ = 2.5 cm	Apple, etc. Mercury
	Bracelets				1.5 × 3 cm	Apple, Mercury
	Phone cases					
	Community social networks					Zèbre
Luxury items – jewels	UHF and HF NFC					
	Lingerie, glasses, shoes, etc.		✓		Fight against counterfeiting All formats	Louboutin, Lise Charmel, Ray-Ban, etc.
	Wines, spirits, LCD tags and corks		✓ ✓			
	Jewelry display cases				Security	

Table 5.1. *Technical range of applications for NFC and its antenna forms*

5.1.4. *Targets/tags for the monitoring of long-distance races*

At 13.56 MHz, "RFID" ISO 15 693 gantries measuring 2.5 m × 3 m are installed at the start and finish points, and all along the route of long-distance races (e.g. marathons or horse races), to read the competitors' numbered bibs, verify their passing of the checkpoints and record the times at which they have passed. When each competitor arrives at the finish line, a race controller analyzes their tags one by one, to ensure there has been no cheating or fraud along the way. This analysis is performed with a small, portable NFC reader (a mobile telephone, of course), which conforms to ISO 21481 – NFC IP2, and thus supports ISO 15 693 (see Figure 5.1).

Figure 5.1. *Routes of marathons and horse races (Photos by Pygmalyon/DAGSYSTEM)*

5.1.5. *Targets/tags for monitoring luxury items*

In the process of production, luxury products (suitcases, bags, shoes, sports equipment, lingerie, wines, spirits, jewelry, etc.) are equipped with integrated circuits working at 13.56 MHz, "NFC A, NFC B or NFC V targets/tags", which have a cryptographic unit (e.g. type AES 256) and pass through RFID ISO 15693 gantries (compatible with NFC IP2) measuring around 70 cm × 50 cm which are installed along the length of the production line for the purpose of industrial SCM (*Supply Chain Management*). Once again, using a small portable NFC reader, conforming to NFC IP2 (and thus supporting ISO 14443 and/or 15693) and easily available on the market (e.g. a mobile phone or a watch) when they pass through customs, the border control services can quickly check the authenticity of the product (its origin, whether it is counterfeit, etc.). Finally, when the products reach the store, the retailer and the end customer are able to do the same thing to check their purchases with a standard mobile phone (see Figure 5.2).

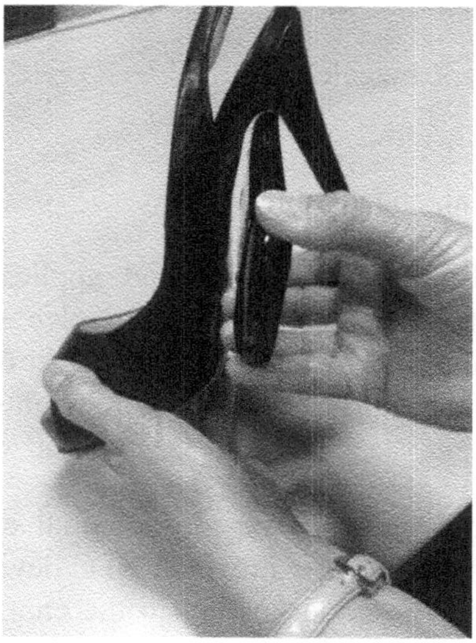

Figure 5.2. *Luxury shoe with an NFC chip (Photo by Tagsys)*

In short, almost anything and everything can now be coupled with NFC. If we wish to be strict, we can construct the "architecture/form" matrix, and add in all cells that the architectures and forms must be "interoperable in the strictest sense", because all NFC elements, in order to be able to claim to be NFC IP1 – ISO 18029 or NFC IP2 – ISO 21481 compliant, must strictly comply with the norms of conformity. There is still some way to go to make this a reality.

5.2. Segmentation, typologies of markets, their problems and their incidences, and direct technical constraints on NFC devices

Let us begin by listing the main applicational segments of NFC at time of writing:

– mobile telephony;

– banking, money matters, payment;

– transport, ticketing;

– consumer electronics (television sets, monitors, refrigerators, washing machines, etc.);

– automobiles;

– wearables (watches, bracelets, etc.);

– medical care and health;

– government, social, and community life.

These all have different typologies and application-related technical issues, and with each of them, we need to solve specific problems and constraints relating to the antennas and numerous other day-to-day problems of integrations, due to the fairly hostile environments in which they are implemented.

We have only briefly touched on the potential existence of numerous sectors of application and markets in which NFC devices have the potential to flourish and bloom. The issue lies not in the enormous number of such devices, but in the vast diversity of types of application for each of them. Many people hope for complete interoperability of this whole wonderful world – covered by the catch-all marketing term "NFC". Such people are in for a rude awakening. For further details on these subjects, see the discussion below.

5.2.1. *Market sectors and typologies*

Indisputably, having an NFC device with a man–machine interface of a screen and/or individual keyboard offers greater flexibility for multiple and varied ideas of application. Obviously, it is well-known that a mobile phone can easily become a sort of Swiss Army knife with a screen or keyboard, serving as a camera, a GPS device, a TV (magnifying glass not included), a Web terminal/computer for those with nimble fingers, a one- or two-cup electric cafetière (well, almost), and sometimes a mobile phone, where the quality is still sometimes average at best as compromises between quality, price, power consumption, and battery life are still so difficult to find. A few years ago, it was entirely normal to build the new up-and-coming technology – NFC – into a mobile phone, and have it emulate a contactless chip card, to perform the functions of money transfers, payment, transport, and so on. How long, then, will we have to wait before mobile phones can make a cup of coffee?

Today, this part of the market – mobile telephony – is only the visible part of the iceberg that is NFC (though undeniably a large one). This market is known, identified and quantified, and this part of the chapter will deal with the problems encountered with these different implementations and detail the integration of the typically NFC part in any NFC device, compliant with IP1 and IP2, and focus mainly on the analog part, software applications aside. Let us state, without prejudice, that we leave all application software architectures to the specialists who, often, do not dare to touch the lower layers of the OSI model. To each his own, as they say.

5.3. Mobile telephony

We shall begin, then, with mobile telephony.

Different architectures can be used for the topological integration of the NFC antenna into mobile phones. Here, then, are a few possible examples of antenna integration:

– in the removable case of a phone, with a connector linking the shell and the phone's electronics;

– on the phone's main system card;

– within the SIM card itself;

– in a removable memory card (e.g. an SD Card);

– on or in the battery;

– integrated into the mechanical part of the case, etc.

The photos (Figures 5.3(a) and (b)) taken from the Website of "iFixit" – a well-known company which repairs mobile phones – give very explicit examples and clearly indicate the position of the antenna (inside the back cover), its size (not exactly identical to the ID1 of chip cards), and therefore its form factor.

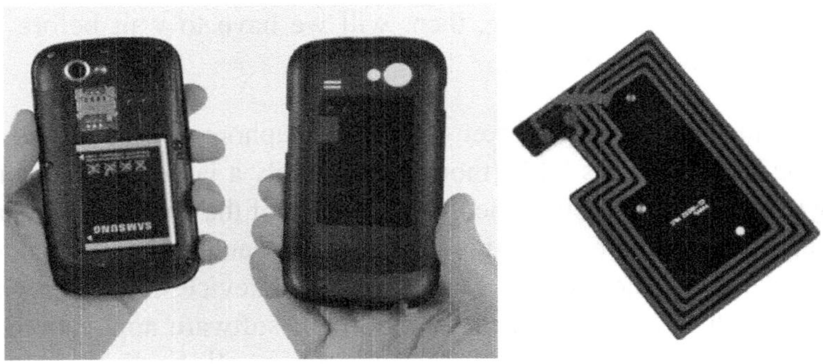

a)

b)

Figure 5.3. *a) Independent antenna within the case of a mobile phone; b) antenna integrated into the battery of a "Nexus S" mobile phone*

Here is another of many possible examples. A few years ago to gain ground in the market, after filing numerous patents, Apple manufactured prototypes of an NFC iPhone, with the fine wires of the 13.56 MHz antenna built into the LCD screen itself.

As readers may realize, the mechanical ingenuity of advertisers, designers, and developers is limitless. Unfortunately, in terms of respecting the values of the magnetic fields produced and ensuring communication in the medium (passive communication, active communication, reader/writer mode, peer-to-peer, card emulation, loading effect, etc.), let us simply and diplomatically say that there is some way still to go, and that this is a different story altogether.

Let us now look at the fields of application and their constraints within which users employ the NFC function of their mobile phone or their NFC device.

Today, this is not a revelation. There are two main fields of application: payment and transport.

5.4. Banks/money matters/payments

This sector of application is, by its very nature, highly sensitive to the security aspects of transactions by means of NFC devices, and great care is taken to verify firstly all the mysteries of the contactless technologies involved, and secondly the security-specific applicative layers and their different implementations (conventional, HCE, Cloud, etc.). Technically speaking, ultimately the PCDs and PICCs (in NFC terminology, the initiators and targets in contactless chip-card emulation mode) and the performances of their antennas in terms of magnetic fields must be made to fit the technical specifications of "EMVCo Contactless",[1] which are specifically oriented toward banks/payments, adapted from ISO 14443 A & B which, remember, are encapsulated by the NFC IP2 norm. Next, the product must undergo all the Common Criteria tests and obtain all certifications such as EAL 5+ and better.

With new eras come new problems. The market of payment and traditional ATMs is a sector evolving rapidly. Indeed, the market for "POS" (*Points of Sale*) – devices that are currently found at all points of purchase (shops, restaurants, etc.) to pay for goods and services with a credit/debit card (contactless or otherwise) – is currently shifting toward smaller, lightweight personal devices, known as "mPOS Contactless" ("m" denoting mobile), and performing the same functions with contactless devices (cards or smart phones) which must also satisfy EMVCo. Additionally, Table 5.2 indicates the similarities and differences between ATM and mPOS (particular attention should be paid to the final ten rows of this table).

1 For EMVCo Contactless, the existing differences are well known for the values of magnetic fields, field volume, tools and testing procedures, etc.

Applications	ATM	mPOS

	Proximity Card	NFC
Contactless norms	ISO 14443	ISO 18092/ISO 21481 (and thus ISO 14443)
Applicable standard	EMVCo*	EMVCo
Contactless device	Smart card	Smartphone
Function	True smart card	NFC – Card emulation
Power supply	*Batteryless*	*Battery-assisted*
Form factor	Known – ID1	Very variable
Antenna category	1–3	–
Reading distance	2–5 cm (1.5 + x)*	1 cm
Field H (A/m)	Strong (4/5)*	Weak (2) (battery)
Coupling coeff. k	Low	High
Coupling index n	Low	High
Loading effect	Weak	Strong
	Dynamic	Static and dynamic
Detuning Initiator	Null – Slight	Great
Detuning Target	Slight	Average
Stack effect	No*	Yes, possibly
Retro-modulation	Average	Weak (PLM) Strong (ALM)

Table 5.2. *Similarities and differences between ATM, POS and mPOS*

5.5. Transport

"Contactless" technology in transport all over the world is based, and relies mainly upon, ISO 14443 part A, with the use of products in the MiFare family of NXP, and partly on part B of the same norm, promoted in France by the STIF and RATP in the region of Paris (application design Calypso and product Navigo) and used in certain cities elsewhere. Of course, it should not be forgotten that in the Far East there is a very high presence of FeliCa products, made by Sony. Furthermore, in that market we also need to take account of the strong interrelations with the municipal and regional authorities, national and foreign rail transport bodies, and others. Expressed thus, technically, it goes without saying that the NFC IP2 norm (which covers the aforecited ISO 14443 A & B and FeliCa) is the solution. Unfortunately, as usual, that would be too easy.

In spite of the fact that no-one wants to redo the work of the ISO, the advent of NFC somewhat changes data, with its variations of NFC IPx, NFC Forum, etc., its NFC devices of all types (form factor, loading effect), its specific protocol between initiators and targets (in-card emulation mode) and the associated timings.

Figure 5.4. *Transport, mobile telephony and NFC card emulation*

While the application to transport of NFC devices – primarily mobile phones (see Figure 5.4) – can boast of being truly NFC, the numerous questions and constraints (both hard and soft) in Table 5.3 remain unresolved, and leave many questions unanswered regarding the definition and design of NFC devices and their antennas.

<table>
<tr><th>Questions</th><th>Impacts and constraints
on the design of the antenna</th></tr>
<tr><td>All the terminals for validating transport tickets and equipment must accept ALL NFC mobile phones (which NFC, IPx?).</td><td rowspan="3">The fundamental basis of the answers to these questions lies in many of the quarters of the circle regarding different form factors, loading effects, etc., which are often not very compatible to ensure true interoperability.</td></tr>
<tr><td>Must we also manage all NFC objects (NFC USB keys, NFC tablets, etc.)?</td></tr>
<tr><td>Is it unrealistic to hope to manage all types of NFC objects?</td></tr>
<tr><td>Is it the responsibility of phone manufacturers to ensure their products are compatible with the validators, or is it the other way around?</td><td>Today, validators are ISO 14443 compliant (hence also NFC IP1 and/or IP2), rather than NFC Forum or EMVCo. The application distances and values of magnetic fields are different.</td></tr>
<tr><td>It must be compatible with the existing range of technologies (validators, current cards, contactless bank cards, etc.).</td><td>Today's validators are ISO 14443 compliant (and thus NFC IP1 and/or IP2), rather than EMVCo, so the antennas need to be updated.</td></tr>
<tr><td>Must the active volume of the field of the validator be modified to accept some or all NFC objects, and does it remain EMVCo compliant?</td><td>The volume of the active field for transport applications is functionally different from the volume for banking applications.</td></tr>
<tr><td>Must we align all the mobiles in hardline card emulation ISO 14443 A & B?</td><td>Obviously it would be ideal, compares to the extensive range of existing validators.</td></tr>
<tr><td>Can we sell the NFC problem from the public perspective, rather than an industrial one, as follows?
"Any coupler/validator must be capable of reading any form of contactless electronic "tickets" – even those in NFC form – regardless of where, at least within the country!"</td><td>This, again, is an excellent issue.</td></tr>
<tr><td>Do we need to write "implementation specifications" and "performance specifications" for ISO 14443 specific to the transport market?</td><td rowspan="3">Under the auspices of the French Transport Ministry (AFIMB), the first-generation European norms CEN TS 16 794 part 1 & 2, specific to contactless NFC transport applications are now available.

They indicate how to manage polling and how to achieve satisfactory user throughput times.</td></tr>
<tr><td>How do we deal with polling and its local variants of the old and new arrivals – A, B and NFC?</td></tr>
<tr><td>How can we stick to the draconian constraints of traveler passage time and data collision management?</td></tr>
</table>

Table 5.3. *Hard and soft constraints remaining to be solved for Transport in NFC*

Given this long and non-exhaustive inventory of issues and constraints to be dealt with, industrialists are now working long and hard on the subject.

Let us again examine a few constraints and technical pitfalls (this book is intended only to be technical; no commercial purpose is intended) with which systems (and their integrated circuits) and their antennas need to deal. In order to do so, we shall examine a few typical "use cases" of this branch in detail.

5.5.1. *1st case: card emulation mode in battery-assisted and then flat battery*

This first application stems from a sentence pronounced by a user: "If I have got in to the metro… I must be able to get out!"… and this certainly makes sense. However, there is a flaw in this statement, which we shall now analyze:

– if I cannot get in, it may be my fault, because my transport ticket is not valid. Alternatively, it might be the fault of my NFC mobile phone (mobile switched off, battery flat, battery removed, mobile or battery broken, etc.), but initially we can assume it is not the transport operator's fault;

– on the other hand, if I have managed to get in, this means that my ticket (my target NFC device in "*card emulation – battery-assisted*" mode) was working properly, and so now if I cannot get out, whatever the reason (for instance, if I have drained the battery, then so be it… I am within my rights to make calls on the subway), then it is the transport operator's fault. Also, if I cannot get out, I may believe that the transporter is holding me against my will, and take them to court. This last point implies, therefore, that the initiator of the exit terminal must be able to provide enough energy to the target, now operating in "*card emulation – battery flat*" mode, to remedy the failure of the battery of "all" types of NFC mobile phones in the world of users (foreign tourists with NFC applications on their own local NFC phones), whatever their form factors, their loading effects and their levels of energy consumption. There is still a way to go.

5.5.2. *2nd case: card emulation mode with battery out*

This case concerns the problems linked to a certain type of fraudsters. As we know, fraud is dreadfully ugly, but it does exist. The protagonist in our scenario may well have jumped the barrier to avoid paying. Yet – horror of horrors – he spots the ticket inspector! He proudly presents his NFC mobile phone, having previously taken care to remove the battery. Knowing, on the one hand, that legally the inspector cannot force anyone to replace the battery in their phone if they have removed it and, on the other, that numerous electronic parts of the mobile phone need to be supplied with power in order to check the validity of the ticket, the transport operator needs to be sure to provide the inspector with a (small and portable) device capable of remotely powering all mobiles on the market (of course including all card emulators with battery out).

In these two cases of application, the magnetic field of an external reader has difficulty in providing all of the remote power necessary for the NFC device to work properly. Hence, for these applications it is crucial to design a clever system to manage the power and the possible placement of a second antenna devoted to providing power to the parts (including encryption) vital to the conditions of access.

5.5.3. *3rd case: collisions and loading effects*

An active young woman, who has two NFC mobile phones – one for her personal life and the other for her professional activities, presents her handbag, unopened, to an NFC validator/initiator (which is perfectly normal… this is "contactless"). In the bag, her personal and professional transport passes are present respectively in the two NFC phones.

Even if we spare readers the "trivial" problems of polling software to manage and resolve data collisions between the two NFC devices in contactless card-emulation mode, the singular selection of the correct phone and the correct transport pass to debit, we still need to deal with the fact that the loading effect is greater with two mobile phones than it is with one. In addition, having to remove the correct mobile from the bag in order to go through the barrier would greatly decrease the fluidity of the contactless transport pass system, and would not be acceptable for the application. There are, as yet, a great many cases to resolve.

5.6. Automobiles

Automobiles have been using contactless devices for a long time, such as access and anti-theft systems (Remote Keyless Entry), hands-free access (Passive Keyless Entry) and hands-free startup (Passive Go), etc. These are essentially proprietary systems, and work mainly in LF at 125 kHz and UHF at 433 and/or 866 MHz.

For many years, keys or badges – often in the mechanical form of a "smart card", contactless or otherwise – have been used to transfer the vehicle's service information to the after-sales team during checkups (odometer reading, distance traveled between two checkups, fuel level, state of charge of the battery, date of checks, repairs, and diagnostics, etc.). Automobile manufacturers have developed contactless chip card systems standardized at 13.56 MHz, compliant with ISO 14443, which are generally able to pay at parking meters and freeway tolls, access private and public parking lots, and also to gain entry to vehicles (private, private fleets or rental vehicles), start them and locate them in a parking garage. The smart card version described above only lacked a screen (for control and viewing), a keyboard and the pinnacle of RF communication techniques. Readers must now realize that an NFC mobile phone that is able to emulate a contactless chip card could (but only *could*) serve this purpose. Let your imaginations run wild!

For example, in 2008, BMW gave a demonstration of its *Connected Key* (see Figure 5.5), equipped with an integrated NFC circuit and a mobile phone demonstrating the easy transfer, between the vehicle and the "ignition key", of content pertaining to navigation information, the vehicle and/or its personalization. In addition, when the key is held close to an NFC mobile phone, the content is automatically displayed on its screen.

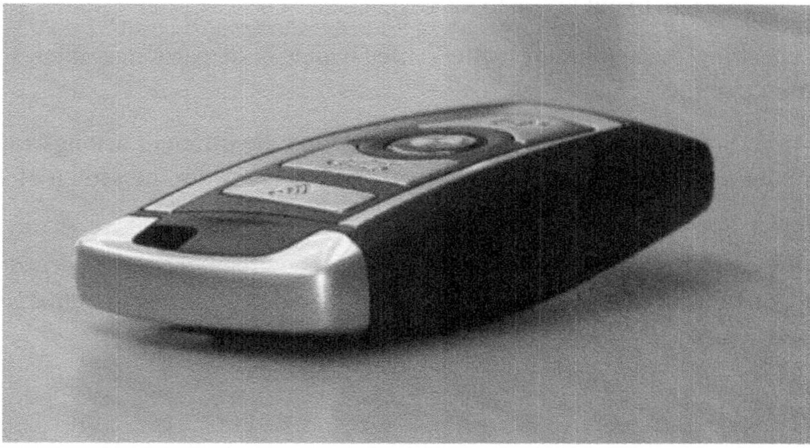

Figure 5.5. *Example of BMW's first "Connected Key"*

NFC technology has made many *urbi et orbi* functions (i.e. inside and outside the car) functions available for keys. Nowadays, the main manufacturers such as Valéo, Continental, Delphi, etc., are focusing their research on the way to dematerialize the key completely and integrate its primary functions into a device that almost all car drivers have to hand: a mobile phone.

In this automobile market, one of the main issues is security against theft. Faced with this situation, these systems employ heavily-secured and authenticated communication in the area of NFC (which exhibits the advantage of only being active over a distance of a few centimeters) and communicate with the car's sensors in less than a tenth of a second, whilst consuming very little energy.

In summary, the "key of the future (including NFC) + mobile phone (including NFC)" combination for a car becomes a source of communication, notably facilitating the following, in addition to the conventional functions of allowing access to the vehicle (opening or closing of the locks):

– authorized driver selection;

– use of the NFC mobile phone as a "virtual ignition key", encrypted and secured to start the vehicle;

– at any time and remotely, checking the state of the vehicle and its location (e.g. using the GPS information of the vehicle and of the mobile phone to find (by password, anti-theft, etc.) where it is parked, and, using the navigation system to guide you to the car);

– checking the remaining battery life, which is of particular interest for electric vehicles;

– from home and/or from the office, to define the different settings for the air-conditioner, radio, seats, programmed startup, etc. from the screen of your mobile;

– the car being able to communicate with its environment and its driver (the GPS position of the vehicle in case of accident, accessing emergency contacts, etc.);

– enabling the users to maintain a link with the manufacturer, for example to download information or applications (background colors for the dashboard screen, etc.) from the manufacturer's website;

– communicating with a tablet or a mobile phone to transfer data from a directory, etc.

Other possibilities, thanks to the key's pure NFC communications, include:

– onboard connectivity;

- quick and easy pairing of onboard equipment and entertainment devices (car radios, loading of audio files, CD players, management of passenger screens),

- maintenance (collection of data and car info),

- access to motor management services,

- automatic recording to security systems, etc;

– transactions;

- telematics,

- secure contactless payment,

- reserving train tickets and hotel rooms,

- connections to carsharing sites, localizing a rental vehicle,

- access to and payment for car parking, purchase of fuel, prepayment of tolls, with the support of the user's phone or directly integrated into the dashboard.

What happens if the key is stolen? It has a system to detect its owner by reading their fingerprints and even their hemoglobin levels.

A few examples are given in Figure 5.6.

5.6.1. *Consumer electronics*

Here again, to illustrate our point, let us take a representative example of consumer electronics applications for NFC, which are presented early every year at the major Consumer Electronics Show (CES) in Las Vegas.

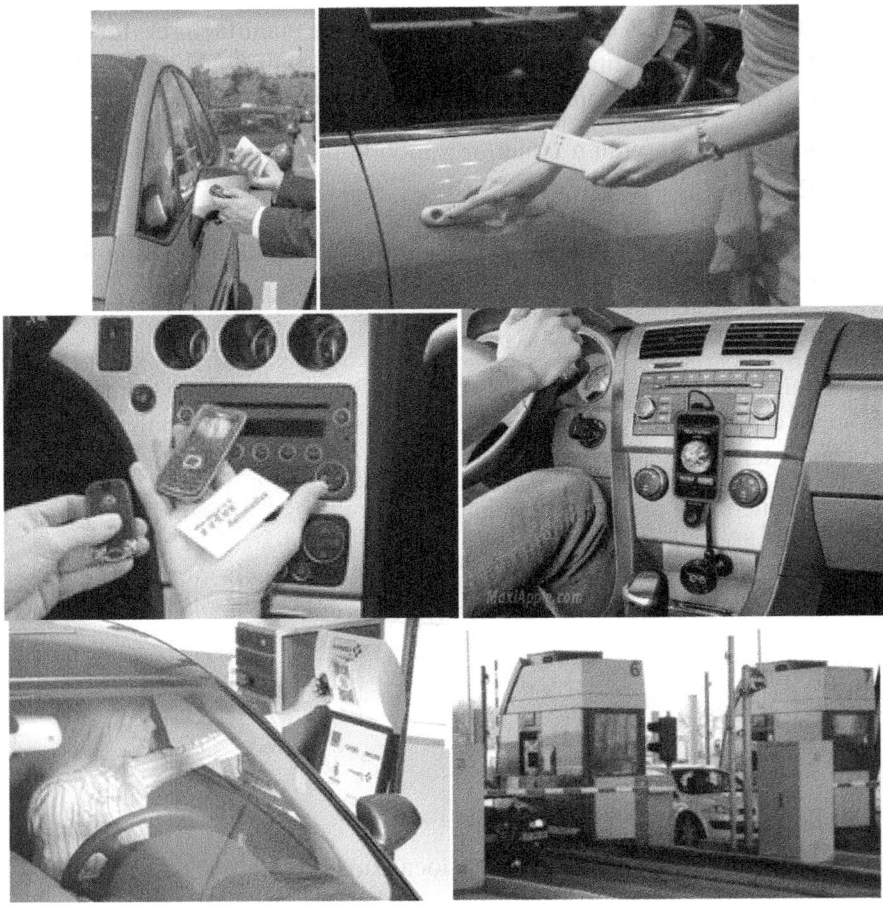

Figure 5.6. *Examples of applications of NFC in the automobile market*

One example is the digital photo frame to display photos of your grandchildren. Granted, it is a photo frame, but it is an NFC frame to download photos directly from your mobile phone. Where on the frame, though, do you present the mobile phone? In which orientation, portrait or landscape? At a particular point? Absolutely anywhere on the surface of the frame? Anywhere around the border of the frame? The form factor and loading effect change completely, as do the applicational ergonomics and the conformity to NFC norms.

a)

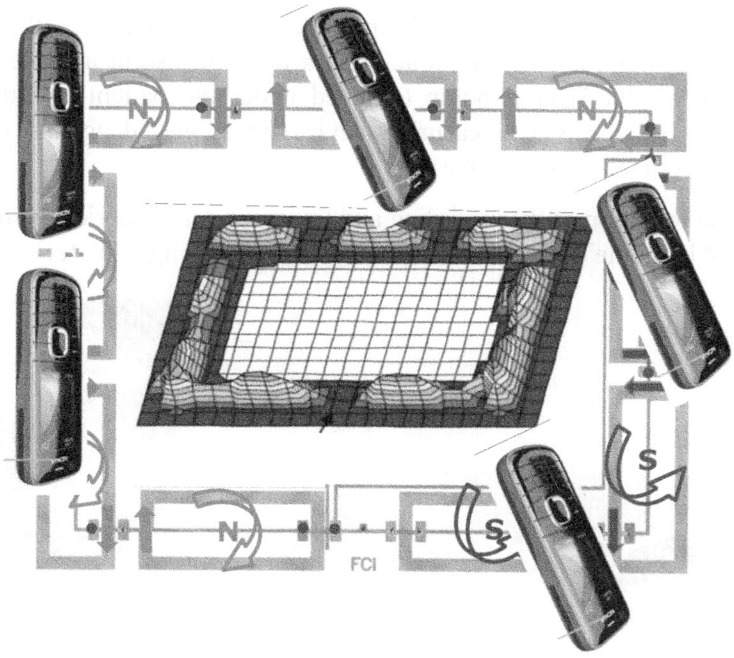

b)

Figure 5.7. *Example of an NFC antenna (patented) arranged all around the frame. For a color version of the figure, see www.iste.co.uk/paret/nfc.zip*

Figures 5.7(a) and (b) show the example of an NFC antenna (patented), arranged all around the frame, and thus with a particular form factor, so as to be able to accept any position and presentation of the mobile phones, having nothing to do with other NFC applications.

The same is true for NFC devices embedded in car radios, stereos, Hi-Fi speakers, television, PCs, etc.

5.7. Healthcare

NFC has also found numerous applications in the medical domain, and professional and personal devices. Let us briefly mention three examples:

– application for measuring blood pressure and monitoring weight (records, treatments, files, tracking/tracing, etc.) which are transferred by NFC to the mobile phone;

– for monitoring diabetes, NFC provides the link between the device measuring blood glucose, recommending a dose of insulin and the control mechanism of the insulin pump (situated under the patient's clothes, for example), with the NFC measuring device being placed against it if the patient agrees (see Figures 5.8(a) and (b));

– the automation of the recording and transmission of services to patients, dispensed by home caregivers with a mobile phone using NFC technology.

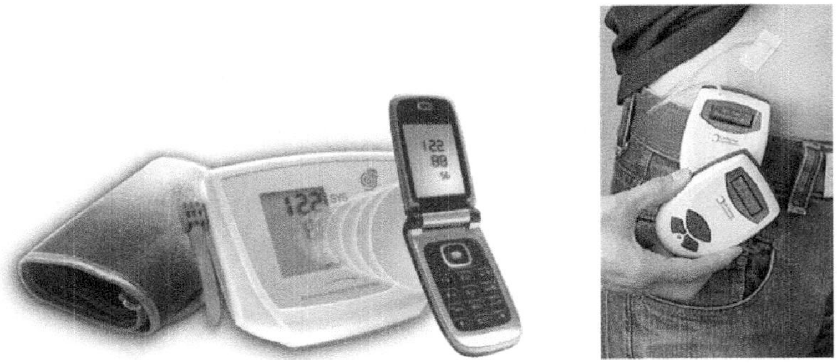

a) b)

Figure 5.8. *Diabetes monitoring with NFC*

In addition, there are hundreds of NFC connected bracelets (see Figure 5.9) which are beginning to appear for our well-being, monitoring our health, to the great relief of our insurance companies who, thanks to Big Data, can raise our health insurance premiums sky high.

Figure 5.9. *NFC connected bracelets*

5.8. Communicating objects

There are multiple applications for NFC, and their fields are highly varied.

The form (and the associated form factors, as well as the antennas) for NFC communicating objects are also very varied, and are in vogue for the coming decade. Let us begin by examining the cases and applicational typologies of simple NFC objects. To clarify the ideas, the aim in the following sections is to present a few non-exhaustive examples (using France as a case study) of the broad range of communicating objects which are beginning to flourish on the international market.

5.8.1. *NFC tablets*

These products are well-known on the market. Some of them may substitute a bank card when making purchases online, but can also be used for numerous other applications (health, access control, delivery of additional information to museum visitors; see Figure 5.10).

Figure 5.10. *Google's NFC tablet*

5.8.2. *NFC USB key*

This NFC device is an "NFC USB key". Another form factor, another loading effect, its shape and its small size mean that it is a pocket-sized product which establishes a link between the world of the Internet (PC, WEB) and NFC products of all norms/RFID, cards, tags, and mobiles in the world of secure transactions. We can cite the concrete example of the application of a perfect tool for topping up a transport pass online in a public environment. All the user needs to do is present the NFC USB key to the validator, if it accepts that form factor different to a card or a mobile phone (see Figure 5.11).

Figure 5.11. *NFC USB stick (NeoWave document)*

5.8.3. *NFC communication and promotional objects*

The widespread dissemination of smartphones, and the advent of objects connected to one another or to the Internet, facilitates the emergence of new ecosystems of services – particularly in communication and advertising. There are multiple uses for these services: identification, access control, payment, ticketing, and tag reading. Beyond these uses, their integration into the different services links to mobility, audio-guiding to a geolocation, increasingly involving social networks, or augmented reality.

A few years ago, the company StrapMedia designed an architecture devoted to tourist and urban contactless services providing high-added-value services by compounding all those aimed at tourists on various contactless supports, particularly in an NFC object (see Figure 5.12):

– a public transport card to get around the city;

– ticketing to grant queue-jump access to museums and attractions;

– an electronic wallet to regulate one's local purchases;

– access to geolocalized or generic multimedia content;

– promotional offers made available based on a client profile.

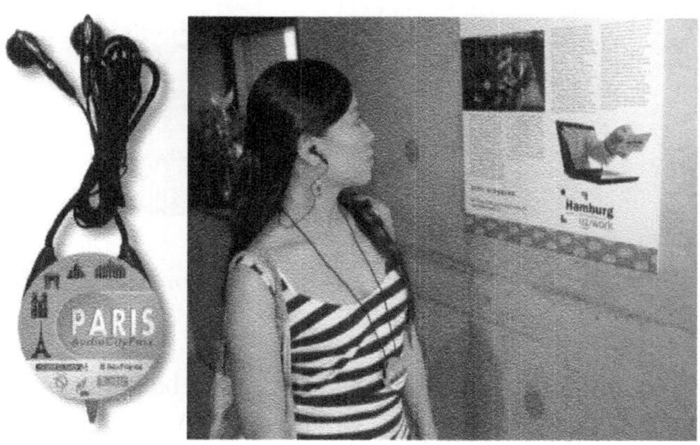

Figure 5.12. *Communicating and publicity objects for NFC (StrapMedia document)*

In addition, the NFC device (again with particular form factor and loading effect, and a new appropriate antenna), mixing contactless services and audio content, is a personalized companion, ideal for all urban tourists and all their applications.

5.8.4. *"Add-on" NFC devices, sticks, and stickers*

5.8.4.1. *External*

Why not render any modern mobile object compatible with NFC technology? This is the basic idea of an "NFC sticker" (see Figure 5.13), which involves sticking to the back of a mobile object an NFC module capable of communicating with the object's internal electronics, via Bluetooth wireless communication. The presence of Bluetooth is standard in many portable items, which automatically ensures compatibility with the vast majority.

The sticker turns the object into an NFC mobile with the twofold function of a card emulator and an NFC reader. It can function both in passive and active mode.

Figure 5.13. *NFC sticker (document from TwinLinx)*

Of course, in order to work, the sticker needs to be battery-assisted and of small dimensions, and includes a small solar/PV panel to recharge the battery, which prevents the battery life of the NFC device from being adversely affected.

5.8.4.2. *Internal*

The "NFC add-on sticker" solution is designed to be added to a mobile device equipped with a SIM card. For example, the solution presented (see Figure 5.14) was designed to be inserted physically between the existing SIM card and its connector, so as to be mechanically and functionally unobtrusive to normal operation, and to superimpose the function of NFC when it is needed.

Figure 5.14. *NFC add-on sticker (document from Mercury Technologies)*

This completes our initial exploration of the main fields of application of NFC. You, the readers, will undoubtedly discover many more during your personal and professional peregrinations to add to this chapter, which will soon become absolutely enormous!

PART 3

Applicational Constraints Needing to be Solved when Designing NFC Systems and their Antennas

Introduction to Part 3

This third part lays down the framework and the context in which we design an NFC device and its antennas, be it an initiator or target, in active or passive mode, batteryless or battery-assisted.

Aside from the design of the antenna itself (for details and calculations, see the other work by this author: *Antenna Design for NFC Devices* [PAR 16], the problems and their direct relations arise mainly at a number of levels, presented in the form of two distinct chapters in order to clearly distinguish the part played by each.

Chapter 6 describes the *structural constraints* linked to the direct and indirect consequences of the intrinsic content of the NFC norms and standards (ISO, NFC Forum, EMV, etc.), and the conditions that must be satisfied in order to be worthy of (read "compliant with", in the judicial and legal sense of the term) the name "NFC" (in active or passive mode, batteryless, or battery-assisted, etc.), rather than being (without wishing to be disparaging) a simple ISO 14443 system. These constraints include:

– the form factors of the antennas;

– the variations in operating distance, and therefore in magnetic coupling;

– the values of the coupling coefficients and indices;

– the values of the retromodulation voltages;

– the values of the quality coefficients;

– the various modes of operation for each slice of time.

Chapter 7 firstly presents the *functional applicational constraints* pertaining to various applications (loading effects of targets, stacks of targets/tags, etc.), followed by the problems of the antenna's direct environment (effects of the cases, metal, presence of a battery, EMC pollution) and how to deal with these constraints. The problems in question include:

– initiators and targets deliberately not tuned and/or accidentally detuned;

– constraints linked to the "stacks";

– environments hostile to wave propagation.

6

Structural Constraints in NFC

As promised in the introduction to Part 3, this chapter describes the structural constraints as a result of the intrinsic content of the NFC protocols (ISO, NFC Forum, EMV, etc.), and the conditions that must be satisfied in order to obtain the name "NFC" (in active or passive mode, batteryless or battery-assisted, etc.). Indeed, all too often because of ignorance or verbal laziness (the short term NFC is so easy to pronounce!), but sometimes because of a deliberate corporate desire to lead their customers astray, some users deliberately mix the terms "ISO 14443 contactless chip card readers" and true "ISO 18092 or 21481 NFC devices", which certainly have many similarities, but also exhibit profound differences, as discussed previously.

6.1. Constraints due to the form factors of the antennas

Let us begin by dealing with the most typical structural constraints.

Structurally, the text of the generic NFC norm, ISO 18092, does not indicate any particular form for an NFC device or the size of their antennas (which can range from the size of an ant to that of an ocean liner or even larger!), nor the operating distance. Each device is free to have its own applications and its own problems. Certainly there are problems to speak of. Essentially, one of the main problems with NFC lies in the definition of NFC devices, initiators, and targets, in terms of their physical aspects, their mechanical forms, and dimensions – simply put the "form factor" – and of course, of the antennas, their forms, and their resulting relative magnetic couplings:

– firstly depending upon the fields of application (mobile telephony, consumer electronics, automobile, etc.);

– secondly, depending on the specific application within the same field (e.g. for different models of mobile in the family of mobile phones – flip-phones, slide-phones, with a fully-integrated keyboard and screen, etc.).

6.2. Constraints due to variations of the operating distance

In HF, at 13.56 MHz the operational principle is of a near (electro-) magnetic field, and thus the devices work by inductive coupling between the antenna of an NFC device (an initiator or target) and another NFC device (again, an initiator or target), in reader, writer, emulation card or peer-to-peer modes, and depends specifically on the distance.

6.2.1. *Distances and magnetic couplings*

Depending on the type of NFC device and its modes of operation and operating distances, the values of magnetic coupling to achieve functionally are not remotely similar to those typically found in standard contactless applications between base stations and conventional plastic chip cards in ID1 format. Indeed, the near electromagnetic environment of the antenna of a standard contactless chip card is generally very light (an antenna with a number of spires sandwiched in plastic). An NFC device such as a mobile phone or a tablet is not at all like this, and its environment (metal shell, battery, numerous integrated circuits, screen, etc.) contributes greatly to the degradation of magnetic coupling [PAR 16].

6.3. Constraint of the maximum acceptable value of the quality coefficient Q of the initiator antenna

As already indicated several times, NFC uplink use ASK m% modulation of the carrier frequency. To ensure the communication functions properly, the maximum value of the overvoltage coefficient Q_max of the initiator antenna must be such that the bandwidth Bp_ant (at -3 dB) equal to f0/Q is at least capable of channeling all of the frequencies contained in the spectrum of the signal modulating the carrier frequency. Thus, in the worst case (corresponding to the maximum value – Q_max – of the initiator antenna circuit), the fundamental bit rate of the square-wave signals (with a cyclic ratio of 50% in the case of NRZ bit coding) of that stream of digital data must be at least equal to half of the bandwidth Bp_ant of the tuned circuit – i.e.

$$2 \times \text{bit\_rate} = \text{Bp\_ant\_min}$$

or indeed:

$$2 \times \text{bit\_rate} = \text{f0/Q\_max}$$

Hence,

$$\text{Q\_max} = \text{f0}/(2 \times \text{bit\_rate})$$

To improve our understanding, we shall discuss a few examples.

EXAMPLE 6.1.–

– carrier frequency:

f0 = 13.56 MHz

– digital data rate transmitted by the initiator:

bit_rate = 106 kbps

– in that case, we find:

Q_max = 64

Unfortunately, as already indicated, for NFC uplink ISO 18092 (or ISO 14443, type A), in order to ensure maximum energy transfer the signal modulating the carrier frequency is far from having NRZ 50/50 bit coding. In fact, it is modified Miller bit coding in which, for a bit duration of 9.44 μs, a “pause” on the carrier frequency of duration Tp = 3 μs is made. That pause Tp is equivalent to the transmission of a frequency whose period is 2 Tp = 6 μs, which is an equivalent bit rate of (1/2 Tp) = 166 kbps, so:

$$\text{Q\_max} = \text{f0}/(2 \times 1/2\ \text{Tp})$$

$$\rightarrow \text{Q\_max} = \text{f0} \times \text{Tp}$$

This means that:

$$Q\_max = 40.68$$

This maximum theoretical value is not acceptable or usable, because it is always necessary to secure systems in all scenarios (tolerances, drifts (L, C, and frequencies), etc.). Suffice to say that in the example above, we could use a Q_max_useable of around 35.

EXAMPLE 6.2.–

– carrier frequency:

$$f0 = 13.56 \text{ MHz}$$

– in the case of applications using NFC IP2 – ISO 21 481 with authorized use of ISO 15693 and NFC–V targets (such as I_code circuit of NXPs), whatever the bit rate, the shortest time present in the uplink communication protocol is a "pause" lasting Tp = 9.44 µs. In this case, we find Q_max = 128, which can generally be reduced to Q_max_usable = 100 for the same reasons as above.

NOTE.– Generally, these values of Q_max_usable are not too difficult to obtain, because the technological designs of the antennas lend them specific Q_max (no-load) values which are much higher than those required for the applications, and are easy to reduce using serial resistors.

This is what happens to the maximum applicative value of the quality coefficient of the initiator antenna. To conclude our discussion of this subject, let us briefly say that the quality coefficient Q is also one of the main parameters influencing the fall and rise of the carrier frequency times (and thus the field produced) during "pauses" in ASK 100% and the rapid variations of modulation in ASK m%. For this purpose, we refer to Chapters 8 and 9, which give a detailed examination of the antenna circuit's pulsed response.

6.4. Constraint of the value of return (retro) modulation voltage

After transmitting ASK commands, the initiator switches to listening mode to monitor for responses from the target. To this end, with the exception of "peer-to-peer" mode, the initiator constantly transmits a pure carrier frequency, and in order to attain understanding waits for the target to indicate its presence and respond by way of a particular modulation, which represents the variation of its load, whether passively (PLM) or actively (ALM).

6.4.1. *Passive Load Modulation (PLM)*

6.4.1.1. *Retromodulation voltage*

By reciprocity of mutual induction between the target and the initiator, the value of the variation in voltage, ΔV1, induced in the coil of the initiator antenna during the return modulation is given by the formula reciprocal to that giving the value of the voltage induced by the initiator in the target antenna, so that:

ΔV1 at the edges of the initiator antenna

$$\Delta V1 = \Delta V2 \times M \times \omega \times I2$$

or indeed:

$$\Delta V1 = \Delta V2 \times Q1 \times k \times \sqrt{(L1/L2)}$$

Obviously, this voltage is lower when the distance between the target and the initiator is greater, because the coupling coefficient and the values of mutual inductances are lower.

EXAMPLE 6.3.– Where:

$L1 = 0.5\mu H$

$L2 = 5\mu H$

$Q1 = Q2 = 30$

$k = 1\%$	if $\Delta V2 = 5$ Vpp, then $\Delta V1 = 500$ mVpp
$= 0.3\%$	$= 150$ mVpp
$= 0.01\%$	$= 5$ mVpp

Of course, the electronic receiver part of the initiator must be capable of handling such signals. On the other hand, when the target is situated very close to the initiator antenna (e.g. a mobile phone pressed against an mPOS reader) the coupling coefficient "k" can be as great as 20–30%. Without accounting for the presence of a shunt or a loading effect, in the same conditions as above we would have obtained:

$$k = 20\% \qquad \text{if } \Delta V2 = 5 \text{ Vpp, then } \Delta V1 = 10 \text{ Vpp}$$

$$= 30\% \qquad\qquad\qquad\qquad = 15 \text{ Vpp}$$

NOTE.– These values need to be significantly corrected due to the loading effect.

This voltage ΔV1 is added to the voltage already present across the initiator antenna. In the resulting signal the carrier frequency is also slightly modulated by that return signal, and then has a modulation index of a few %. That signal, present across the initiator antenna, is also retransmitted, with its own specific spectrum and thus its own side bands, which must, of course, fit into the FCC and ETSI templates.

6.4.1.2. *Following which…*

Having glimpsed the major question of the approximate value of the induced voltage recovered, we must now deal with a whole host of new problems, which are also very important:

– how do we amplify this minuscule voltage (hundreds of millivolts, short-distance and proximity applications), superimposed on the tens of volts that are already present on the initiator antenna during the interrogation phase;

– what is the value of this voltage in relation to the disturbing "noise signals", such as industrial noise picked up by the antenna (e.g. radiation from PCs), and the values of the noise signal specific to the base station's own amplifier;

– what is the overall estimation of the signal-to-noise ratio (SNR) of the setup?

Nor should we forget the problems inherent to the specific characteristics and qualities of the signals present in cases where multiple transponders are present in the magnetic field. Let us now consider the concerns.

6.4.1.3. *Concerns*

While respecting the RF regulations mentioned in the previous chapters, with devices conforming to ISO 18092 or ISO 14443 using chip cards in the classic mechanical format ID1 (45 × 85 mm), a reading/operational range of ~ 5-10 cm can be achieved. Generally speaking, the factors limiting this range of communication are:

1) having the ability to provide sufficient energy (remote power) to the contactless target to function within the range of the initiator's magnetic field/power;

2) having good reception in both forms and amplitudes of the data transmitted by the initiator;

3) the possibility for the initiator to receive the data from the target, which requires sufficient magnetic coupling between the initiator antenna and the target antenna (or chip-card emulator) with known values of the coupling factor k and mutual inductance M;

4) knowing that the operating range decreases considerably when very small antennas are used (on contactless SIM cards and micro SD cards used in a mobile phone or a tablet, for example);

5) additional barriers between the antennas, placed either deliberately or accidentally – e.g. the metal layers or screens of the battery and the cases – soon lead to communication problems, so the target may no longer be accessible to an external initiator.

In order to solve this problem, we must examine how to eliminate these limiting factors one by one.

In the case of the range of energy provision, this problem can be resolved very easily. We merely need to locally provide the necessary energy to the contactless target with an auxiliary battery, turning it into a "battery-assisted NFC device". Such is the typical case of a mobile phone which already has its own power supply (when the battery is not flat) to power the peripheral devices, such as a micro SD card, SE card or SIM card). Care must be taken to properly study all the different cases of batteries: *"ON"*, *"Low"*, *"OFF"*, *"flat = discharge = 0"* and *"OUT"* (even more new vocabulary), and deal with them using particular antenna solutions.

In this case, the problem of the transmission of return data from the target to the initiator, is more complex to solve. Indeed, even when we have a solution with a battery-assisted target using a local power source and use of PLM, this does not prove to be a good solution, because it provides only a tiny amount of power in comparison to a conventional solution (passive target + PLM) unless the magnetic coupling is significantly improved. Thus, we still need to find a better solution.

6.4.2. *Active Load Modulation (ALM)*

In addition to the sections already presented on this subject, to resolve the above, one possible solution is to use a different method to PLM to generate a signal presenting the same spectral characteristics as the PLM signal and transmit it actively back to the initiator, in what is known as *Active Load Modulation* (ALM) mode. Notably, this is one of the preferred methods to use with small targets (e.g. a watch, bracelet, NFC USB key, etc.), and so in the case of weak coupling and devices with small batteries included.

Returning to a technical perspective, in the case of the NFC 18092 norm (and also with ISO 14443), during the return communication the overall structure of the frequency spectrum results from the superposition of the unmodulated carrier frequency sine wave sent by the initiator and the modulated return signal from the target (whether in PLM or ALM). Hence, it is composed of the line at 13.56 MHz due to the carrier frequency, and two additional lines corresponding to the two subcarrier frequencies, which are arranged symmetrically around the carrier frequency signal at + and – 848 kHz (thus at 12.712 MHz and 14.408 MHz). The dynamic content of their associated modulation side bands, on both sides, make up the overall content of the return channel (see the shapes of encoded return signals, using MCS or BPSK, in the previous chapters).

In order to transmit the content of the data to the initiator from the target whose spectrum and energy are exclusively contained in the modulations of the side bands of the two subcarrier frequencies above, it is necessary to produce a signal whose spectrum is appropriate for that purpose (as has been done for a long time in UHF RFID in ISO 18000-6). Because of the "half-duplex" transmission principle used by the initiator–target couple, and given that during the time period corresponding to the modulation of the return communication, the carrier frequency of 13.56 MHz is not modulated and is transmitted constantly by the initiator, it does not need to be generated and transmitted by the target. Therefore, it is only necessary to create and transmit – i.e. in active mode (whether in battery-assisted or batteryless application) – a strictly synchronous "Active Load Modulation (ALM)" signal, of known phase, comprising only the content of the signals of the two modulated side bands.

A simple and well-known principle for creating this type of signal with the above properties is "Double Side Band (DSB) modulation", also known as "suppressed carrier" modulation, which can be easily done completely independently of whether the target is battery-assisted or batteryless, by using a ring modulator circuit, suppressing the transmission of the carrier frequency and keeping only the side bands. This type of amplitude-modulated analog signal can be obtained by multiplying two sine signals of different frequencies:

$$U\_mod = U1 \bullet \sin(\omega 1 \bullet t) \bullet U2 \bullet \sin(\omega 2 \bullet t)$$

In this case, as the incident signals present are binary (high/low states), DSB modulation of the signal can be obtained much more easily using a simple ASK modulator, acting as a ring modulator, because the multiplication of the signals is equivalent (in binary) to ASK modulation and can be achieved by a simple "AND" Boolean logical operation (see Figure 6.1).

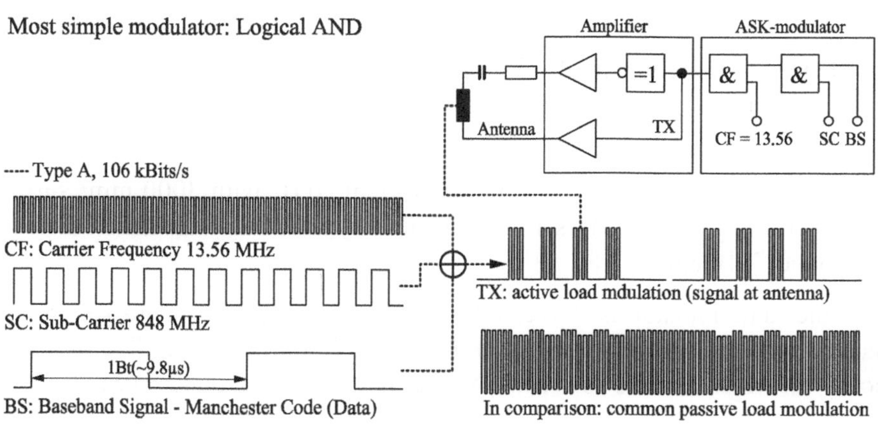

Figure 6.1. *DBS modulation in ALM mode*

The input signals to be given to the modulator are:

– a carrier-frequency signal of 13.56 MHz (fc), issued by and synchronous with the initiator's incident carrier frequency;

– a subcarrier frequency (SC), created locally by division of the carrier frequency;

– a baseband data signal (BS) reflecting the information content which the target wishes to send back to the initiator.

The output signal from the ASK modulator is similar to that of a ring modulator, and provides the required DSB signal. The amplitude of this signal is then brought up to the required level by an amplifier (which often implies that the target be battery-assisted, but it is also possible in batteryless mode), and it is then broadcast by the target antenna (see Figure 6.2).

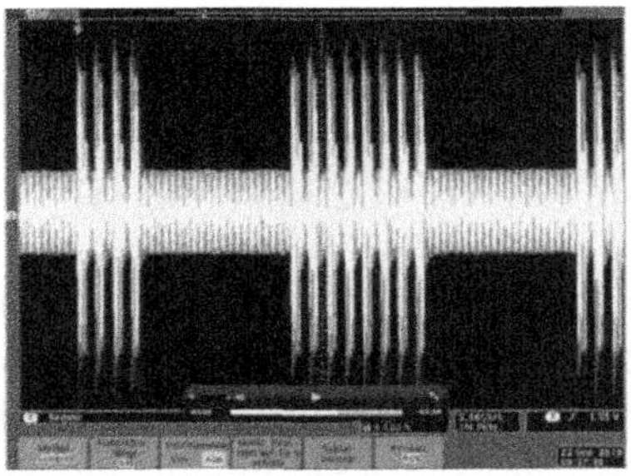

Figure 6.2. *ALM signal broadcast by the target antenna*

Hence, in the context of the use of the ISO 14443 type A or NFC ISO 18092 norms:

– a conventional initiator equipped with an antenna ~ 8 × 8 cm and a target with an antenna in conventional format (ID1, with 3000 mm^2 surface area), working with PLM, is able to obtain a typical communication distance of around 7–10 cm;

– the same initiator as above, still with the same size of antenna (8 × 8 cm), using a target whose antenna is still in the classic format (ID1, 3000 mm^2) but now with ALM, enables us to achieve a reading range of around 50 cm;

– using an initiator whose antenna is no larger than 10% of the surface area of an ID1 contactless card (around 300 mm^2 – e.g. 18 mm × 16 mm) and an ID1 target with a surface area of 3,000 mm^2, operating in ALM, it is still possible to obtain a communication range of around 25 cm. By reducing the surface area of the initiator antenna by a factor of 10, we halve the reading range, which corresponds approximately to the cube root of the ratio of the surfaces between the antennas of the two initiators. This technology is therefore particularly well suited for delivering an acceptable operating range, even with an initiator which has a very small antenna.

EXAMPLE 6.4.– An initiator whose antenna is the size of a micro-SD card is capable of an operational range in air of around 10 cm. Even if that initiator antenna (in reader mode, for example) is installed in a mobile phone whose environment is particularly hostile (with a metal casing, battery, additional shielding, etc.), an acceptable communication range of a few centimeters (3–5 cm) can still be achieved.

Generally, a special RF interface (battery-assisted), linked to the secure element, is required to create the modulation signal necessary to increase the reading range and to handle the very weak signals received by the initiator antenna, owing to its small size.

7

Functional Applicational Constraints

In the interest of completeness, in this chapter we discuss two major domains which are not structurally linked to the principles of NFC *per se*, but are found in day-to-day functional applications as a result of the proliferation of contactless chip cards, RFID, and of course, NFC. These are primarily constraints of "tuning/detuning" of antennas and management of stacks of initiator/target antennas.

7.1. Antenna tuning/detuning constraints

The tuning and detuning (or "offsets") – deliberate or accidental – of the initiator and target antennas are extremely complex issues. One could fill an entire book with a discussion of this topic alone, and still not cover it in sufficient detail. For simplicity's sake, let us take a few examples.

7.1.1. The *case of an initiator*

By its very principle, the initiator's purpose is to transmit and function at 13.56 MHz, and for that to be the case everything must be properly tuned. However:

– when the distance between the initiator and target becomes smaller, the magnetic coupling (static and dynamic) becomes greater, and the loading effect inevitably alters the tuning of the initiator;

– when, during its operation, the NFC device changes mode – e.g. the initiator reverses its role and becomes a target (such is the case with a mobile phone, switching from reader mode to card-emulation mode) – the deliberate tuning of the initiator undergoes a phase of intentional detuning (intentional offset) in target mode.

In order to understand this in better detail, let us take a look at some electronics.

7.1.2. *Electrical representation of the target*

Consider the equivalent diagram of the HF part of a target (see Figure 7.1), which shows the resistance Rp (the equivalent of the input resistance of the integrated circuit Ric, arranged in parallel with the resistance R2 of the antenna).

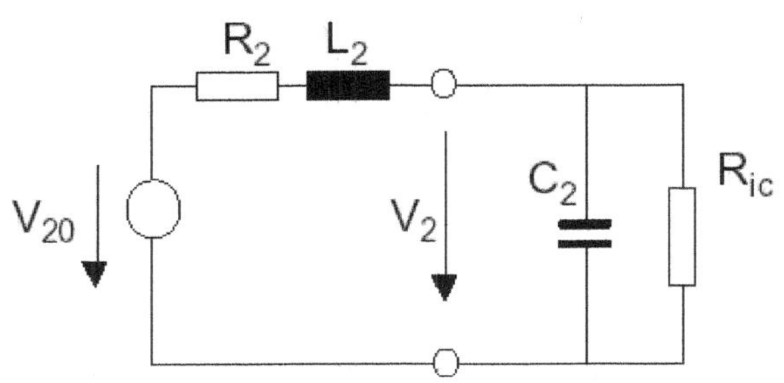

Figure 7.1. *Equivalent diagram of the HF part of a target*

The general equation giving the voltage v_ic (v_2) present at the terminals of the integrated circuit is then written as follows (// meaning "parallel"):

$$v\_ic = \frac{(Rp//Cp)}{Z\_L_2p+(Rp//Cp)} \times v20$$

Following expansion and reduction, with the voltage induced by the initiator v20 being replaced by its value (– j M ω I1), we obtain:

$$v\_ic = \frac{1}{\left(1-L_2p\ Cp\ \omega^2\right)+j\left(L_2p\ \omega / Rp\right)} \times (-jM\ \omega I1)$$

From the value of the complex variable represented by the result of this expression in equation form, if we set (L2p Cp ω r²) = 1, we can deduce the value of the modulus of V_ic:

$$v\_ic = \left(\frac{M \times \omega \times I1}{\left[\left[1 - \left(\frac{\omega}{\omega r} \right)^2 \right]^2 + \left[\frac{L_2 p \times \omega}{Rp} \right]^2 \right]^{1/2}} \right)$$

V_ic is proportional to I1 and, again, as we know that (M × I1) = N2 × Φ21 = B_d × N2 × s2 = μ × H_d × N2 × s2, we obtain:

$$v\_ic = \left(\frac{\omega \times (\mu \times H\_d \times N2 \times s2)}{\left[\left[1 - \left(\frac{\omega}{\omega r} \right)^2 \right]^2 + \left[\frac{L_2 p \times \omega}{Rp} \right]^2 \right]^{1/2}} \right)$$

All else being equal, V_ic is proportional to H_d.

7.1.2.1. *Minimum threshold magnetic field*

The value of the magnetic field H_d (thus for a distance "d") as a function of the voltage V_ic (or vice versa), is, therefore:

$$H\_d = \frac{\left[\left[1 - \left(\frac{\omega}{\omega r} \right)^2 \right]^2 + \left[\frac{L_2 p \times \omega}{Rp} \right]^2 \right]^{1/2}}{\omega \times \mu \times N2 \times s2} \times V\_ic$$

In order for a target's integrated circuit to work properly and serve its purpose, the voltage V_ic must reach at least the minimum voltage V_ic_min stated and guaranteed by the integrated circuit manufacturer, regardless of whether it is tuned to the value of the carrier frequency transmitted by the initiator. If we introduce Qp2 = L2p ω/Rp into the equation, we see that it is achieved for a value of the magnetic field known as the threshold H_d_t ("_d_t" denotes *t* hreshold at a distance *d*):

$$H\_d\_t = \frac{\left[\left[1-\left(\frac{\omega}{\omega r}\right)^2\right]^2 + \left[\frac{1}{Qp2}\right]^2\right]^{1/2}}{\omega \times \mu \times N2 \times s2} \times V\_ic\_min$$

An example of the shape of the variations of H_d_t as a function of ω (or f) for a target integrated circuit with an imposed minimum voltage V_ic_min is shown in Figure 7.2.

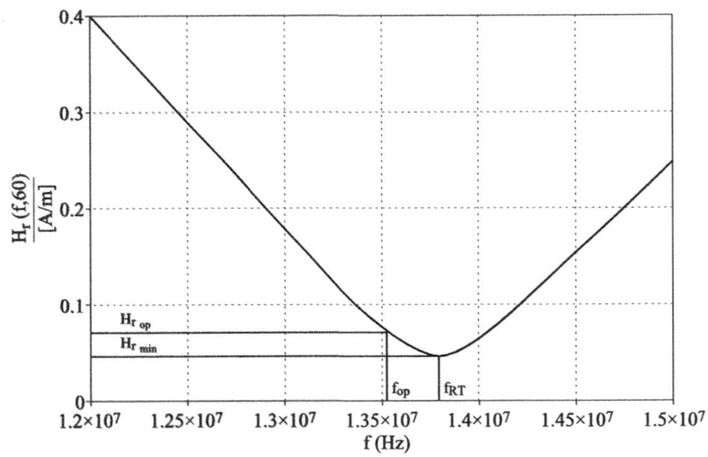

Figure 7.2. *Example of variations of H_d_t as a function of ω (or f)*

Where the target is tuned to the carrier frequency, ω c = ω r, the equation for the threshold magnetic field takes the following simple form:

$$H\_d\_t = \frac{1}{\omega \times \mu \times N2 \times s2\ Qp2} V\_ic\_min$$

It should be noted that, because of the hyperbolic shape of the equation H_d_t = f(Qp2), it is somewhat pointless to try to increase the value of the quality coefficient Qp2 in order to greatly decrease the value of H_d_t.

7.1.2.2. *Stacked targets*

Numerous applications involve targets which can be "stacked" on one another. A simple example of a stack in NFC can be found in the case where several targets (contactless public transport cards, office access cards etc.) are side by side in a wallet, or their positions are slightly staggered. This also applies to two mobile phones (e.g. one for work and one personal), pressed together at the bottom of a bag when passing through a gate to access the office with an ISO 14443 or NFC-compatible device.

7.1.2.3. *The off-center tuning technique*

The theory of coupled circuits applied to stacked target antennas indicates that all the target antennas interact with one another, causing each other to become detuned from their individual initial settings, and that the values of each of their tuning frequency and the overall frequency of the stack tends to decrease. Hence, in such situations it is useful to deliberately tune each target, one by one, to a frequency around 20–40% higher than the carrier frequency – a technique known as *off-center tuning*: for instance, with a system operating at 13.56 MHz it is not uncommon to deliberately tune a target to around 1.25 times the carrier frequency – i.e. 16.95 MHz (see Figure 7.3).

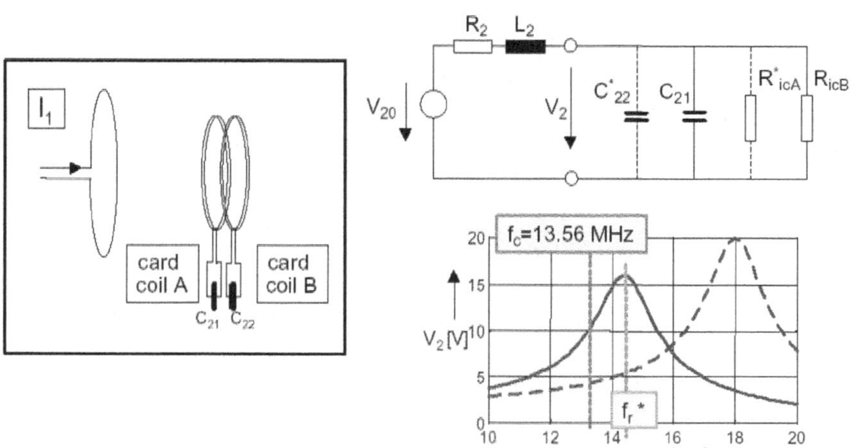

Figure 7.3. *Detuning and the off-center tuning technique*

In view of this principle, and given that sooner or later, many targets will likely be stacked in a wallet or a bag, they are not (or should not be) tuned exactly to the carrier frequency employed by an initiator. Devices which do not allow for these scenarios will inevitably encounter issues. Obviously, this has repercussions for the value of threshold magnetic field/induction (H_d_t, B_d_t) necessary for correct operation and, consequently, for the operating distance of each of the individual targets and that of all the targets when stacked.

A target designed to use the off-center tuning technique has of course a higher threshold magnetic field than a target tuned to the carrier frequency (look again at Figure 7.2). For example, all other things being equal, a target tuned to 16.95 MHz may require around 1.2 A/m, as opposed to 0.2 A/m, to work correctly, as compared to the same elements tuned precisely to the carrier frequency (13.56 MHz).

7.1.2.3.1. Example

As an example, let us examine the case where the initiator works with a carrier frequency equal to fc = 13.56 MHz, and produces a field H_0 = ~ 4 A/m r.m.s., which equates to induction B_0 = μ H_0 = 5 μT r.m.s. Here, we shall use an old integrated circuit whose working threshold voltage is V_ic_min_= 4 V r.m.s (independently of the value of the working carrier frequency).

To begin with, consider the case of a target T0 in ID1 format whose antenna has N = 4 spires measuring 85 × 45 mm, Q = 13.5 and whose tuning frequency *ft* is equal to the carrier frequency – that is, ft = fc = 13.56 MHz. In this case, its threshold field H_d_t @ 13.56 MHz is equal to:

$$H\_d\_t = \frac{1}{\omega \times \mu \times N2 \times s2\ Qp2} \times V\_ic\_\,min$$

$$= 4/((4 \times 3.14\ 10^{-7}) \times (4 \times 38.56\ 10^{-4}) \times 13.7 \times (85.157\ 10^{6}))$$

$$= 177\ \text{mA/m r.m.s}$$

$$B = \mu\ H \text{ or } \mu = 4 \times 3.14\ 10^{-7}$$

so:

$$B = 222.3\ \text{nT r.m.s at ft} = 13.56\ \text{MHz}$$

Target			T0	T1	T2
Working frequency of the initiator	fc	MHz	13.56		
Tuning frequency of the target	ft	MHz	13.56	16	19
ft/fc ratio	ft/fc		1	1.18	1.40
Field supplied at d = 0 by the initiator	H_0	A/m r.m.s	4		
Induction supplied at d = 0 by the initiator	B_0	μT r.m.s	5		
Threshold induction for V_ic_min = 4 V r.m.s	B_d_t	nT r.m.s	222	825	1507
$(B\_0/B\_d\_t) = (1 + a^2)^{3/2}$	B_0/B_d_t		22.5	6	3.3
Thus: a = d/r, approximately	a		2.63	1.65	1.25
Hence, the operating distance, considering that r = 7.9 cm, is	d	cm	21	12.6	6.3

Table 7.1. *Effect of detuning on the threshold induction and field*

Let us now examine what happens when, at the same inductance value of the target antenna (L2 = constant, so Q2 = constant also), for the sake of argument we deliberately "detune" its tuning frequency, by modifying the value of the tuning capacity (though in reality, with a given integrated circuit – which has a given input capacity – all we can do is to alter the value of the antenna's inductance – see Table 7.1).

Consider:

– target T1 ft = 16 MHz, so B_d_t@ 16 MHz = 825 nT r.m.s;

– target T2 ft = 19 MHz, so B_d_t@ 19 MHz = 1507 nT r.m.s.

NOTE.– This table is, of course, applicable to all calculations regarding the influence of tolerances, dispersions, etc. of the target's tuning frequency values as a function of the tolerances and dispersions of the components.

Given these numerical examples – with the same strength of magnetic field H_0 produced by the initiator antenna – for a target considered in isolation, we can easily estimate the difference in operating distance between a target tuned strictly to the carrier frequency and a target detuned deliberately, designed for use in a stack, for example. In addition, this table very clearly indicates that in order to work properly with an initiator whose carrier frequency is 13.56 MHz, as compared to a target tuned strictly to the carrier frequency, the target T1 tuned to 16 MHz requires a minimum magnetic induction B of 825nT, which is 825/222 = around 4 times greater (400%) and for T2, ~6.8 times greater. Clearly, with the same level of energy provided by the initiator, this technique of off-center tuning reduces the operating distance and yet, seen from an outside perspective, the target in isolation and the stacked targets appear to have practically identical working distances. The suspense is unbearable!

First conclusion

As we have just seen, with the same initiator, the operating distance of a deliberately-detuned target is reduced in comparison to an optimized target.

Let us now take the (carefully-chosen) example of the particular target T1 operating solo, so up to a distance of 12 cm, and place it in a stack with a few similar devices, with performances strictly identical to that of T1. Once they are stacked, because of the phenomenon of mutual induction between all the antennas present, as shown by Figure 7.4, the entire set is detuned and the "overall" tuning frequency of the stack decreases.

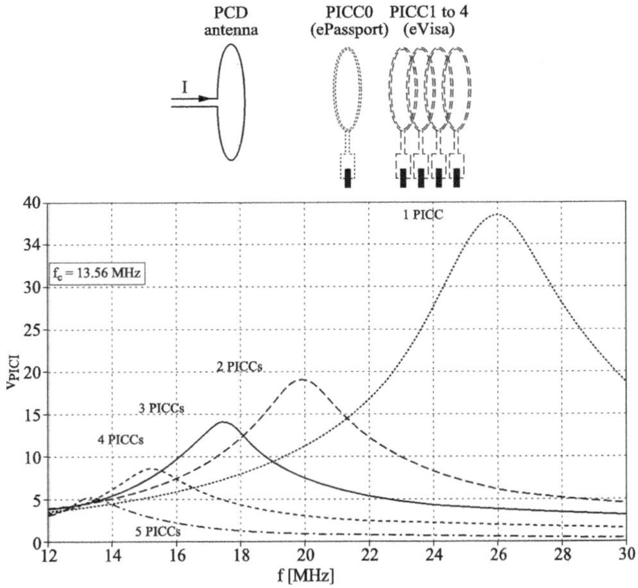

Figure 7.4. *Variation of detuning as a function of the number of targets*

Indeed, despite the fact that the set of targets present in the magnetic field absorbs more energy (so normally, the operating distance of each target should decrease), because the "overall tuning" of the stack approaches the value of the carrier frequency, their individual sensitivities increase, by virtue of the fact that the minimum value of the operating threshold H_d_t is reduced and therefore the operating distance can increase.

Thus, when multiple stacked targets are simultaneously within the field of the initiator antenna, the overall tuning of all the targets tend to come back toward the carrier frequency, and the value of the threshold magnetic field of each of the targets decreases, which, in principle, tends to increase the distance. However, this is practically false, because the effective distance of the initiator supplying energy to all the targets present in the field is then reduced, which, ultimately, makes users believe that an isolated target works at approximately the same distance as those which are stacked. In conclusion, from a user's point of view, all of this is wonderful, because a target taken in isolation will work at ~12 cm, and it is practically the same when multiple targets are stacked as long as each of them is detuned individually initially to an appropriate value (to a frequency slightly higher than the carrier frequency used).

REMARKS.–

– thus, because of the bathtub-like form of the curve H = f(f) (see Figure 7.2), we individually set the tuning frequency of each card higher than the carrier frequency, rather than lower, and act so that the final convergence value of the mathematical series as a function of the initial offset value and the maximum number of targets is very close to the value of the carrier frequency.

– filtering of the return signal when received by the initiator. The values of the subcarrier frequencies of the return link (target → initiator) are, by their very nature, symmetrical in relation to the value of the carrier frequency (see Figure 7.5).

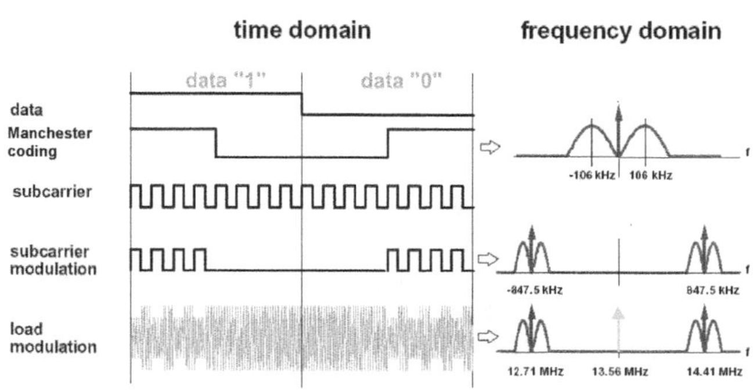

Figure 7.5. *Subcarrier frequencies of the return link, target to initiator*

Because of the detuning of the target tuning circuit, their amplitudes are non-identical, and the tuning value of the input filtering components (see Figure 7.6). thus needs to be adjusted so as to be able to extract the maximum amount of energy contained in the two subcarrier frequencies.

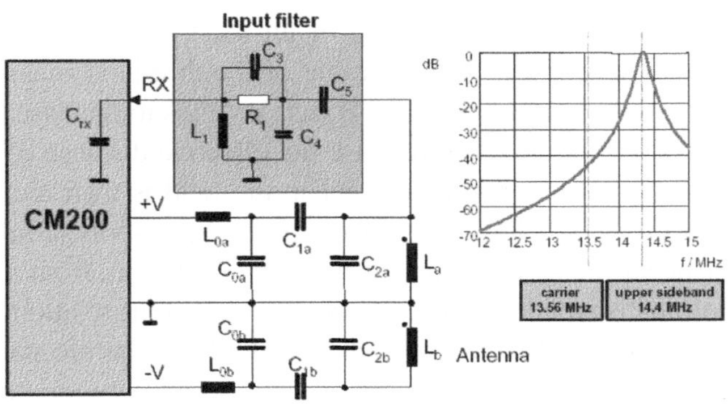

Figure 7.6. *Input filtering of the return signal*

7.1.3. *"Tuned", "untuned" and "detuned" mobile phones*

Throughout the foregoing paragraphs, we have touched on the fact that sometimes it is not desirable to tune the target to the value of the carrier frequency broadcast by the initiator. Having opened the door to this question, we must now resolve it by showing that the value of the inductance of the target antenna (and thus its number of spires, etc.) does actually have an influence. In order to do so, let us briefly examine the main possible applications for an NFC device when used as a smartphone, and discuss the problem which may be encountered.

1) Very frequently, the simplest application of an NFC mobile phone is as a target in card-emulation mode, (what everybody typically uses such devices for (see Figure 7.7)).

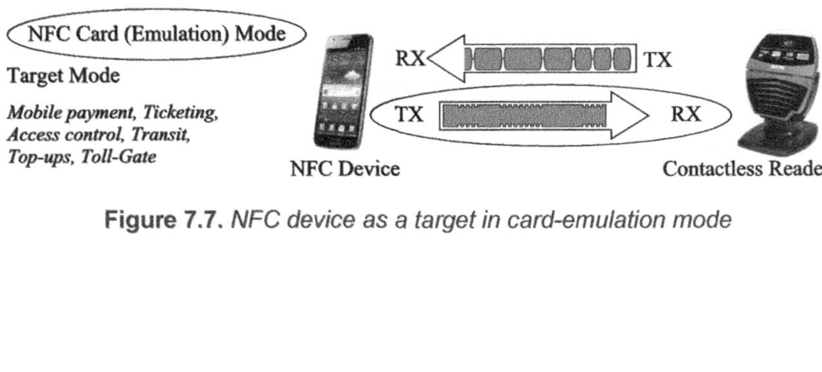

Figure 7.7. *NFC device as a target in card-emulation mode*

Two scenarios may be presented:

a) the mobile phone is a target in card-emulation mode, and is intended to be alone within the field. In this case, refer to Table 7.2;

Mobile phone *acting as a target* in card-emulation, *mono*-target mode	
The initiator interrogating the target mobile phone is in "reader" mode – conventional ISO 14443.	The mobile phone acting as a *mono*-target is in card-emulation mode.
Its transmission is calibrated to 13.56 MHz.	As detailed above, the target should not be tuned to 13.56 MHz, but *deliberately slightly detuned* to a higher frequency (e.g. 14.5 MHz). Hence, the reception antenna of the mobile in card-emulation mode needs to be tuned to this latter frequency, and we need to adjust the filter of the tuned receiver circuit so as to obtain the best possible compromise of the maximum of the higher subcarrier frequency received.

Table 7.2. *Target in card-emulation mode and alone in the field*

b) the mobile phone is a target in card-emulation mode, but its designer has tried to manage the possible effect of stacking two mobile phones side by side in a bag or pocket, in which case there are *multiple* targets present in the field.

Mobile phone *acting as a target* in card-emulation, *multi*-target mode	
The initiator interrogating the target mobile phone is in "reader" mode – conventional ISO 14443.	The mobile phone in a multi-target situation is in card-emulation mode.
Its transmission is calibrated to 13.56 MHz.	Obviously, to allow for the eventuality of stacking of targets, each target is not tuned to 13.56 MHz, and it is desirable for each of them to be *deliberately detuned by some considerable degree* to a higher frequency (17–19 MHz). Thus, we need to differently tune the mobile's receiver antenna to the chosen frequency, and adjust the filter of the tuned receiver circuit differently as a function of the number of targets, so as to obtain the best possible compromise of the maximum of the higher subcarrier frequency received.

Table 7.3. *Multiple targets in card-emulation mode in the field*

2) Shortly afterwards, to serve a different application, the same mobile phone which was a target becomes an initiator, which can function either in reader mode (say, to read a poster on the subway), or in peer-to-peer mode, to create a link between the mobile phone and another device such as a digital photo frame:

a) in reader mode (see Figure 7.8);

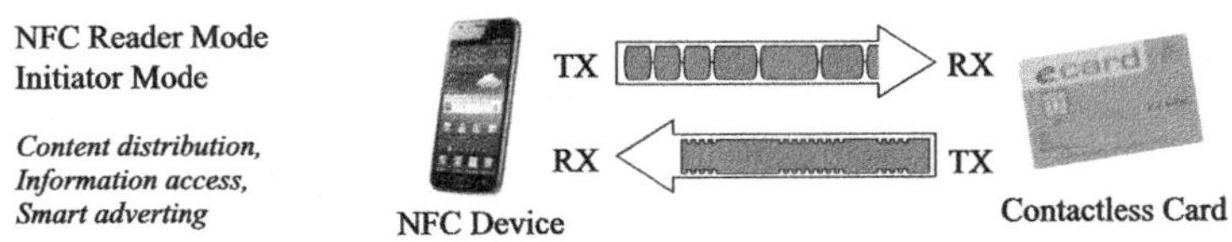

Figure 7.8. *NFC device acting as an initiator in reader mode*

The mobile phone *acting as an initiator* in *reader* mode	
Its tuning circuit for transmission is calibrated to 13.56 MHz.	As the components of its tuning circuit are soldered into place and therefore not adjustable, the mobile phone is calibrated to 13.56 MHz for reception as well.

Table 7.4. *Initiator in reader mode*

b) in "peer-to-peer" mode (see Figure 7.9).

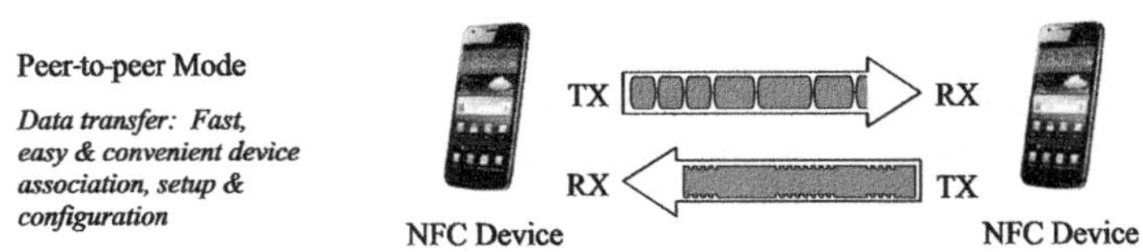

Figure 7.9. *NFC device acting as an initiator in peer-to-peer mode*

The two mobile phones are in *peer-to-peer* mode and operate in *active mode*	
Mobile phone A is first a transmitting initiator and then a receiving target.	Mobile phone B is first a receiving target and then a transmitting initiator.
Its transmission is calibrated to 13.56 MHz.	In *receiver* position, the *receiving target* should not be exactly tuned to 13.56 MHz, but to a slightly higher frequency. Then, the mobile phone becomes a *transmitting initiator*, and it needs to be tuned exactly to 13.56 MHz.

Table 7.5. *Initiator in peer-to-peer mode*

NOTE.–

1) Mobile phones, for financial reasons, are always in a monostatic antenna configuration, i.e. with a *single antenna* for both transmission and reception of the return channel (see Figure 7.10, which represents the conventional setup (integrated circuit, passive components for filtering, tuning and impedance matching, connections, etc.) between the output Tx and the input Rx of the integrated NFC circuits in relation to the antenna which must be installed in the NFC device).

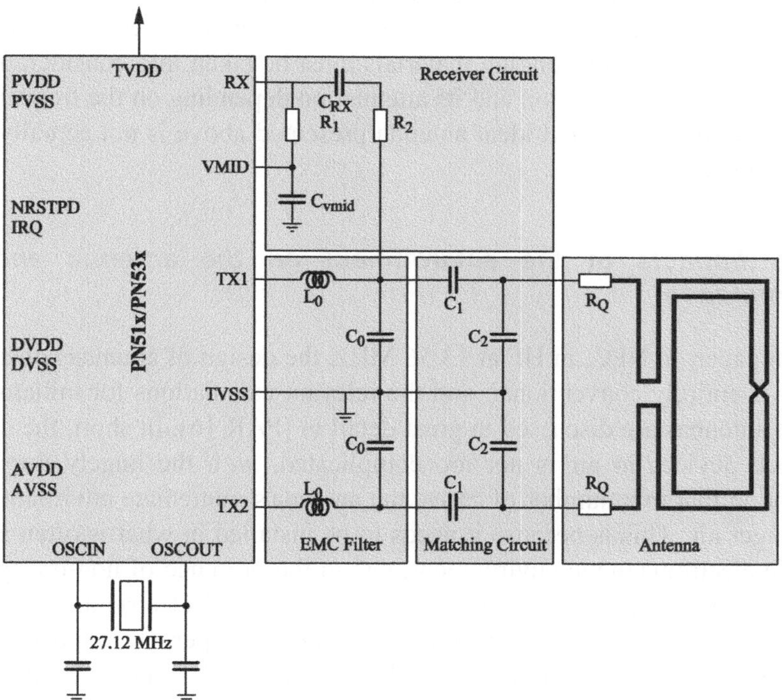

Figure 7.10. *Monostatic antenna configuration*

2) In this setup, once the tuning components of that monostatic antenna are soldered into place, they are permanently fixed, and the values can no longer change, whether the NFC device is in initiator, reader/writer, peer-to-peer, or card-emulation target mode. Given that these modes of operation are profoundly antinomic, it is necessary to find a compromise between all these modes or carefully choose to omit certain ones.

7.2. Constraints and influences of the environment

The aim in the second section of this chapter is to describe the influence of the surrounding electromagnetic environment on the RF electronic component and the antenna of an NFC device, and put forward some possible solutions.

7.2.1. *Physical influence of the environment*

In numerous applications, the transmission of signals unfortunately does not take place solely in air, and numerous environmental parameters resulting from the surrounding materials must be taken into consideration in terms of signal propagation and its attenuation depending on the frequencies. Therefore, the theoretical ideal antenna presented above is not actually very easy to create.

7.2.2. *Impacts of the environment on the antenna and its performance*

On paper, in NFC, in HF at 13.56 MHz, the design of antennas operating in air is entirely conventional, and the relevant calculations for initiator and target antennas are discussed in great detail in [PAR 16]. In short, the design of such devices in air is not too complicated, with the hugely important exception that in a number of cases, the antenna's immediate environment is no longer air. This is because it needs to be installed in what is often a very hostile electromagnetic environment, due to the presence of non-transparent structures which are non-neutral towards the radiated electric and magnetic fields (metal parts made of iron, aluminum, copper, etc.), presence of batteries, the effect of the user's hand and so on, and in addition, all of the resulting loading effects must be taken into account.

This requires us to find a solution so that the electric and magnetic fields (E and H respectively) produced by the antenna of the NFC initiator do not interfere with the electronics of the set formed by the NFC initiator and its environment.

Let us now briefly examine separately the effects and consequences of the environment on the two electrical (E) and magnetic (H) components of the EM field produced.

7.2.3. *Electrical fields E*

At 13.56 MHz, generally the electrical fields E produced by the NFC initiator antenna give rise to RF pollution (e.g. radiation on the clocks of local microcontrollers) and, owing to the types of carrier-frequency signal modulation and a number of nonlinearities of the signals, these fields E also contain high-ranked harmonics, of 65, 69, 70, 135, to 146, which can interfere with the receiver electronics in mobile phones – primarily in the reception bands of GSM850, EGSM900, DCS1800, and PCS1900 RX.

7.2.3.1. *Symmetrical rendering using a compensation-loop antenna*

In certain applications, the leading surface of the initiator antenna is electrically asymmetrical, in which case the antenna has a *hot* point (the active signal) and a *cold* point, linked to the ground, delivering a radiated signal that is asymmetrical in relation to the body. By capacitive coupling, this configuration induces the circulation of currents between the different parts of the antenna and the ground. In order to compensate for these parasitic and unwanted currents, it is necessary to render the setup as symmetrical as possible in terms of its ability not to radiate the electrical field E. Figure 7.11 shows how to obtain the desired compensation conventionally, with the appearance of a new antenna open spire – known as a compensation loop – connected to nothing but which, due to its particular magnetic coupling to the initial active antenna, develops a potential difference of equal amplitude but with the opposite sign.

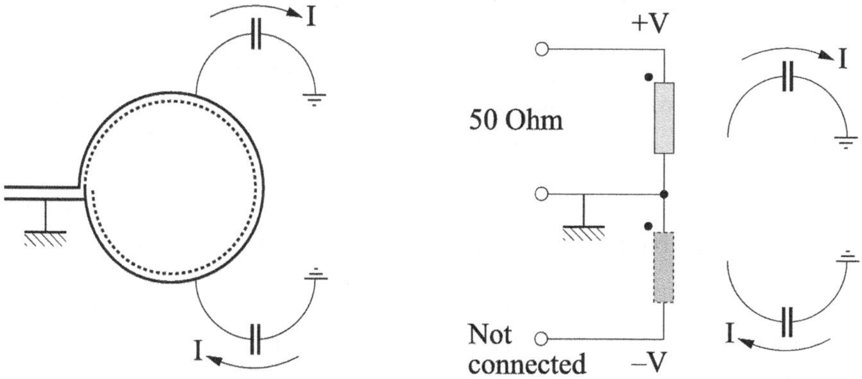

Figure 7.11. *Symmetrical rendering using a compensation-loop antenna*

7.2.3.2. *Symmetrical rendering using a "balun" transformer or using the electronic architecture*

A symmetrical attack by the initiator antenna can easily be achieved, either by using a so-called "balun" (*balanced–unbalanced*) (auto) transformer, which has the ability to transform the asymmetrical attack of the setup into a symmetrical output, or by directly using integrated circuits that approach the antenna network in differential mode. For practical reasons (such as volume and weight) and financial ones, the majority of the NFC market uses the latter solution, which is available from all major component manufacturers.

7.2.3.3. *Antennas shielded by body plate*

It is also wise to try to minimize pollution by placing a shield (linked to the ground) between the antenna and the other electronic or electromagnetic elements in the initiator. This can be done using:

– heavy plates of the printed circuits in the NFC device, appropriately positioned;

– thin metal films (copper, aluminum foil, etc.);

– metal plates;

– the casing of the mobile itself.

To be as effective as possible, the shield needs to be placed as close as possible to the antenna to limit the effects of all the radiated electrical fields.

IMPORTANT NOTE.– The purpose of the shield is to minimize the radiation of the electrical field E, while impacting very little on the magnetic field H necessary for the application. Note that the form of the shield should not cover the whole surface of the antenna, and above all must not appear as a short-circuit spire for the coil of the antenna producing the field H, in which case the system would no longer work at all, but this unfortunate situation is often seen in practice.

Figure 7.12 clearly shows (in the shaded area) what needs to be done so as not to completely cover the surface of the active spires in the antenna with the shielded area, and also leaving the option open to connect the shield to the most appropriate body. Thus, large plates with high mass are to be avoided. It should also be noted that the presence of the shield slightly alters the value of the antenna's inductance.

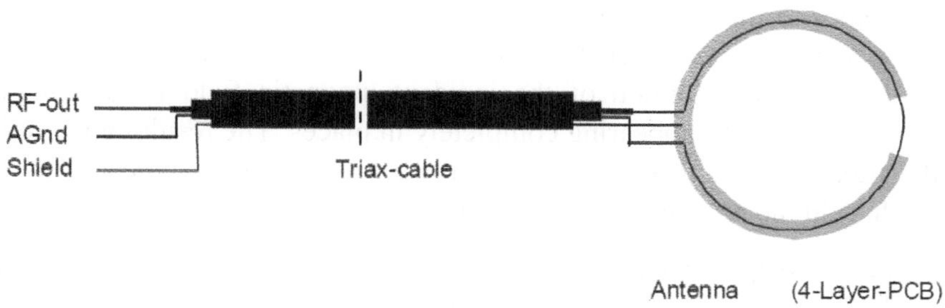

Figure 7.12. *Antenna shielded by a heavy plate*

Obviously this has a cost, and associated drawbacks. Indeed, if the metal shield is placed very close to the antenna, the magnetic field H produced by the antenna gives rise – by induction B – to Foucault currents, better known as "eddy currents" in the metal (see below).

7.2.3.4. *Metal shields and Eddy currents induced*

Owing to Faraday's law, the field and the magnetic inductions caused by the initiator antenna generate, in the surrounding metal surfaces, closed loops of electrical currents which circulate in planes perpendicular to the magnetic field that created them. These currents are known as "Eddy currents" – see Figure 7.13.

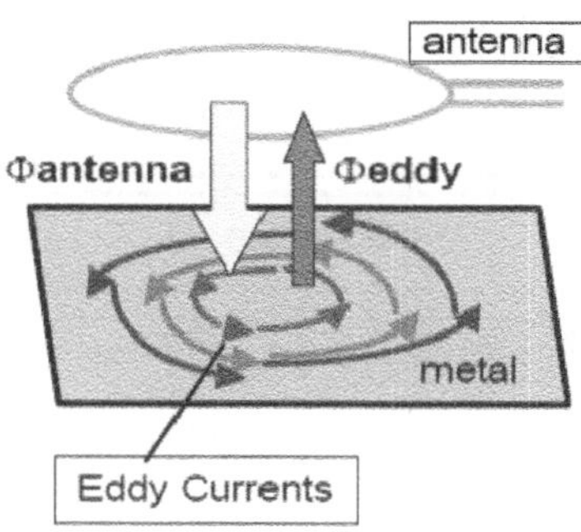

Figure 7.13. *Eddy currents*

The circulation of these closed-loop currents in the metal depends on the geometric shape of the metal plate, the surface in question and, like little loop antennas, these currents in turn create magnetic fluxes ("Eddy fluxes") in the opposite direction (Lenz's law) to those which generated them, and produce a destructive magnetic effect in the opposite direction leading to changes in the distribution of the incident magnetic field lines, with the energy sometimes disappearing completely in places. The resulting reduction of the flux (original flux produced by the reader antenna, decreased by the Eddy flux) in the antenna's coil (the reader or card):

1) alters the apparent value of its inductance which, all other things being equal, leads to the detuning of the antenna;

2) reduces its quality factor, because of the increased watt losses.

7.2.4. *Magnetic field*

If we want the antenna to work properly in the presence of the new metal environment, it is once again necessary to insulate the metal shield, this time using plastoferrite, whose low reluctance controls the magnetic field lines as far as possible. Figure 7.14 illustrates the way in which the ferrite shield reduces the flux circulating in the metal plate, and thus greatly reduces the eddy currents circulating in the metal.

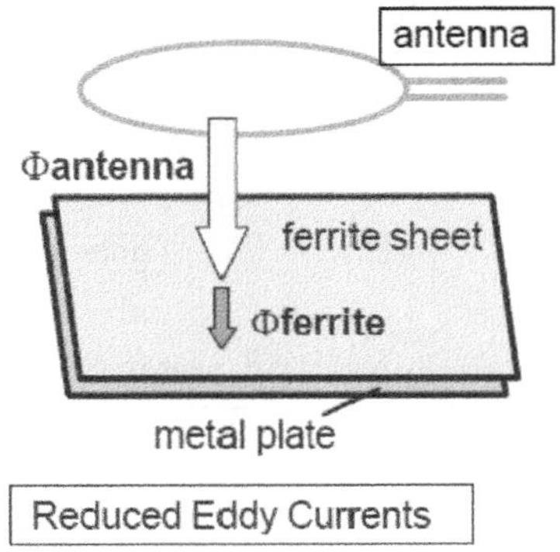

Figure 7.14. *Magnetic shield of plastoferrite*

The presence of this plastoferrite reduces the destructive reflection of power, and produces a sort of counter-reaction effect (as the path of the magnetic field lines H in the ferrite is better known, channeled, and controlled), which means we can more easily characterize the elements present and ensure better reproductivity of the antenna's electrical values in its environment. The detuning of the antenna then becomes more predictable (and can hence be compensated for at the design stage). The metal environment has practically no direct influence on the antenna's electrical values, which mitigates the harmful influence of the loading effect within the element at an additional (but necessary) cost.

7.2.4.1. *Ferrite*

Generally this ferrite shield is industrially made of a thin sheet (100–200 µm) of plastoferrite (which can be easily cut to whatever shape is required), which exhibits good magnetic qualities at the working frequency (i.e. good values of relative permeability μ_r' and especially μ_r'' at that frequency – see later), and finally is very thin so as not to have a major impact on the design of the NFC device, (whose antenna must be able to fit easily into the case, the battery, etc).

It is always possible to improve the quality of that shield by increasing the thickness of the sheet of ferrite, or leaving a little space (air) between the coil of the antenna and the ferrite. Increasing this thickness is the best way to improve performance and obtain greater communication distances, but it is costly.

7.2.4.1.1. Conclusion

In summary, the ferrite sheet serves to shield the antenna largely magnetically against the influence of the metal used to shield the electrical fields and/or conversely against the loading effects. It also helps to protect against the environment (metal part of the casing, influence of the battery, materials, mass, etc.) partly and from the radiation pollution specific to the microcontrollers of the NFC mobile, etc. Yet in all cases, we must bear in mind that the effects and their associated remedies mentioned above mean that it is not possible to use similar communication distances to those obtained in air (with no particular environment to take into account). In a hostile environment, it is typical to find a 20–30% loss of operating distance.

In order to use a longer operating distance, it is sometimes useful to design antennas with specific characteristics of its coils (shared wires, Litz wires, divided spires, distributed spires, etc.) so as to concentrate and/or adapt the value and form of the magnetic field produced. A number of (very expensive) professional tools (such as those made by ANSOFT, AGILENT, etc.)

are capable, after lengthy modeling, of simulating the forms and intensities of the magnetic fields as a function of the coils, arrangements, and environment of the antennas.

7.2.4.2. *Characteristics and magnetic properties of ferrite*

An inductance in air in the absence of ferrite is purely reactive (a pure inductance without watt losses) and, with all else being equal, in the presence of ferrite, we write *as a function of the working frequency* the losses (e.g. through heating) of the set formed by the "coil + ferrite" setup. This indicates that the whole of this new set must be considered not to be reactive only. In order to take account of the physical observations as a function of the working frequency and to be able to integrate them into an equivalent model of the setup (necessary when performing the calculations for external components), it is necessary to introduce into this setup a serial (or parallel) resistor, representing the watt losses of the setup so that the impedance can be written in conventional form:

$$Zs = Rs + j\, Ls\, \omega$$

In order for this equation to be satisfied, we must consider that the permeability μ_r of the material is assimilated to a complex value (a function of the frequency), written as:

$$\mu\_r\_complex = \mu\_r' - j\, \mu\_r''$$

where,

μ_r' = real part (function of ω)

μ_r''= imaginary part (function of ω) which reflects the magnetic losses of the material.

In the above equation, the "–" sign may seem odd, but is necessary in order to be able to formulate the following relation. Indeed, generally speaking, in the context of an equivalent desired representation, we can set:

$$Zs = j\, [Lo\, (\mu\_r\_complex)]\, \omega$$

$$Zs = j\, [Lo\, (\mu\_r' - j\, \mu\_r'')]\, \omega$$

$$Zs = (Lo\, \mu\_r''\omega) + j\, (Lo\, \mu\_r'\, \omega)$$

and, by identifying each term, in turn, we obtain the following for the influence of the frequency on the ferrite material:

$$Rs = Lo\ \mu\_r''\ \omega$$

$$Ls = Lo\ \mu\_r'$$

and,

$$Qs = Ls\ \omega/Rs = \mu\_r'/\mu\_r''$$

7.2.4.3. *Choice of ferrite*

When choosing the ferrite to use, there are a few important physical points that must be considered:

a) the values of μ_r' and μ_r'' are of crucial importance in determining the electrical aspect of the impedance value of the inductor (e.g. constant permeability μ_r' between the values of 40 and 70, up to 10 MHz) as shown above;

b) it is best that the ratio μ_r'/μ_r'' be as high as possible at the working frequency (e.g. 40/1 to 70/1).

Figure 7.15 shows the example of variations in the relative magnetic permeabilities μ_r' and μ_r'' of ferrite IRJ 04 from TDK as a function of the frequency (which is also used as a "flame retardant", in accordance with the flammability norm UL94V-0, which may be of interest for export to the United States).

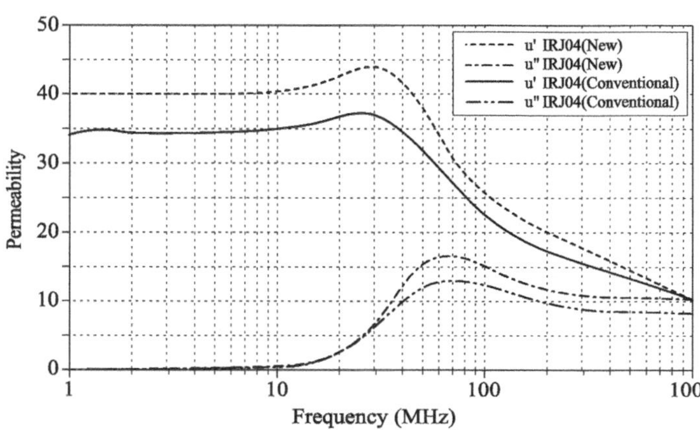

Figure 7.15. *Variations of the magnetic permeabilities μ_r' and μ_r'' as a function of the frequency*

At 13.56 MHz the ratio μ_r''/μ_r' is less than 0.02, demonstrating that the material is a very good magnetic conductor with very slight losses. The usage range of the permeabilities of these ferrites means that they can be used for shielding applications;

c) another interesting specification of plastoferrite lies in the fact that the surface resistivity of the material is high, ensuring good electrical insulation between the ferrite and the other electronic parts of the setup, because the ferrite is often in direct contact with these components (e.g. the resistivity of IRJ04 is greater than 1 MΩ/square inch);

d) one final point relates to the availability of different thicknesses, widths, and lengths of thin, flexible sheets of plastoferrite, with which it is possible to perform a number of tests to validate the solution.

Figure 7.16 shows the non-exhaustive example of flexible and compatible sheets of plastoferrite, made by *TDK*, called "*Flexield*" (IFL 04 & IFL04AB with adhesive tape).

Standard material characteristic / Specification

Material name	IRL04	
Operating temperature range (°C)	-40 to +85	
Initial permeability μ_i (at 10MHz)	35 typ.	
Surface resistivity (Ω/□)	10k min.	
Heat conductivity (W/mk)	1.4	
Standard sheet size (mm)	300 x 200	
Standard sheet thickness (mm)	0.25	0.5
Standard sheet mass (g)	50	100
Density (g/cm³)	3.3	
Environmental friendly	RoHS* & halogen-free	

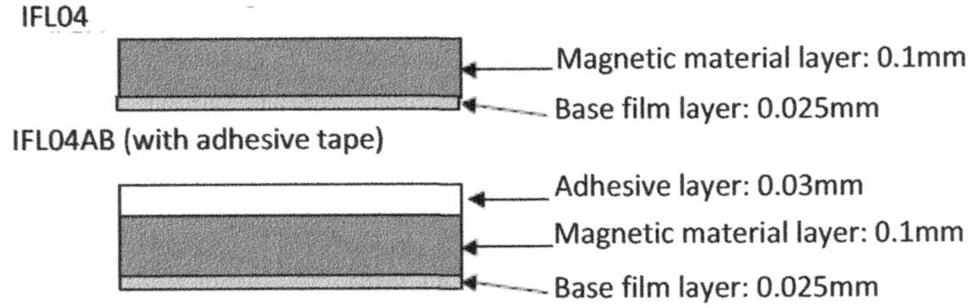

Figure 7.16. *Non-exhaustive example of flexible plastoferrite layers*

7.2.4.4. *Measurements and results*

With a view to ensuring a better understanding of an NFC device, Table 7.6, showing the relevant measurements, absolutely must be filled in when using a metal shielding and/or ferrite plate.

Measuring frequency	Figures	Antenna always retuned to 13.56 MHz				
Type of ferrite:		L	detuning	Q	losses	remarks
Antenna alone	7.17		0			
Antenna + metal shield	7.18					
Antenna + ferrite	7.19		200 kHz			
Antenna + ferrite + metal shield	7.20					
All of the above, in application						

Table 7.6. *Table of measurements to be carried out*

NOTE.– The presence of ferrite generally:

– increases the inductance value, so we need to readjust the value of the tuning capacity (normally reducing it a little because L has increased);

– reduces Q, ensuring a greater degree of damping, which must be taken into account to readjust the external serial resistance for the optimum applicational value of Q (normally reducing it a little, because the R of the coil has increased).

Here are a few examples (see Figures 7.17 to 7.23).

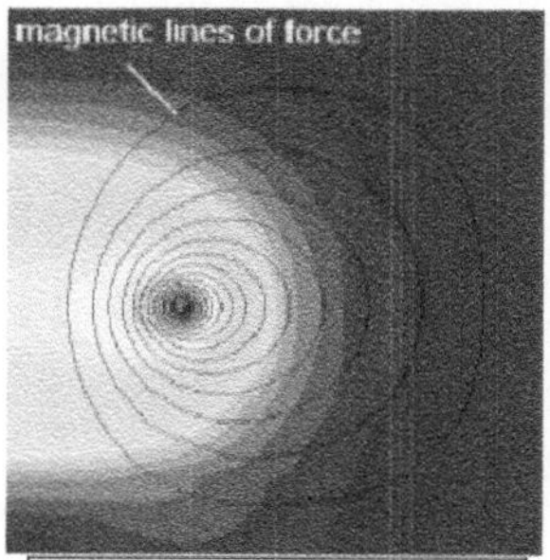

One-turn circular antenna
in free air

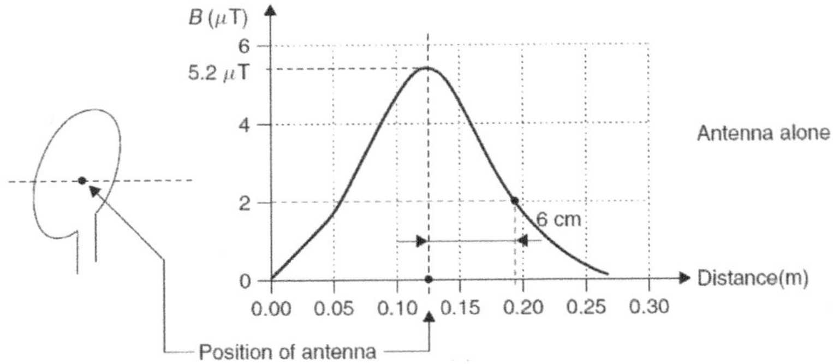

Figure 7.17. *Antenna in air, with no shielding or ferrite plate. For a color version of the figure, see www.iste.co.uk/paret/nfc.zip*

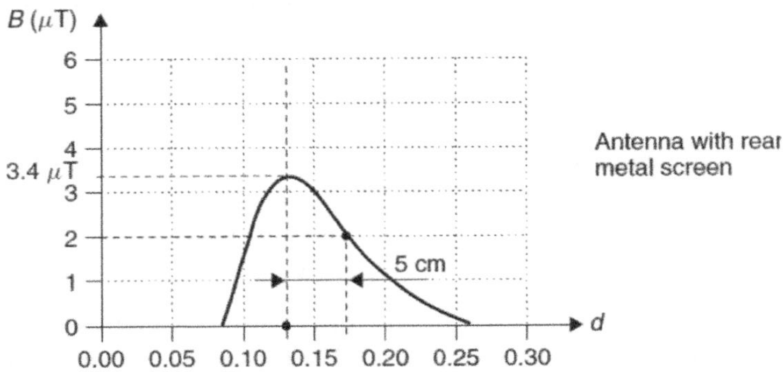

Figure 7.18. *Variety with metal shielding behind the antenna*

Figure 7.19. *Variety with ferrite behind the antenna*

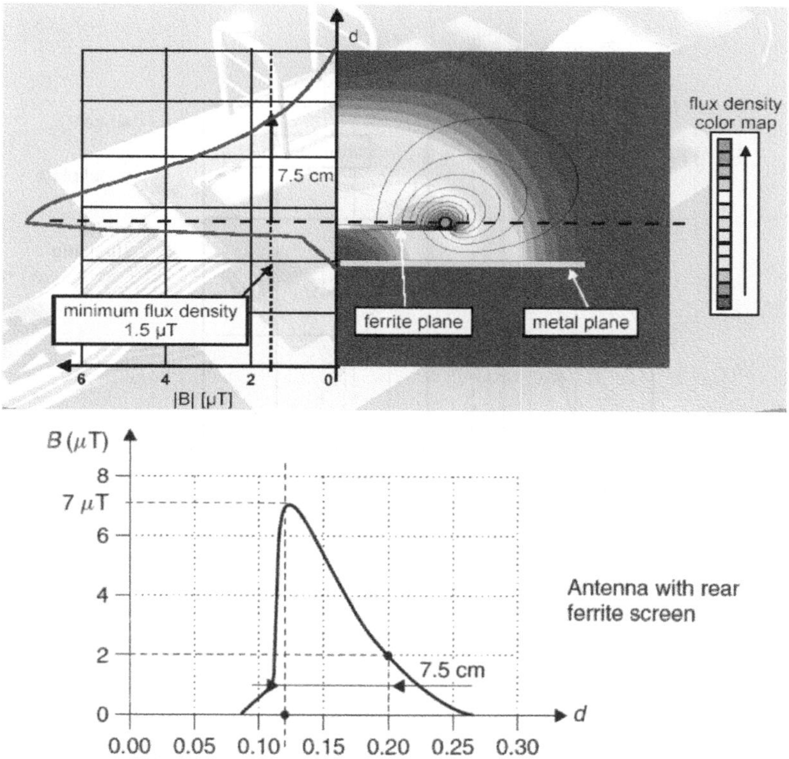

Figure 7.20. *Variety with rear ferrite screen. For a color version of the figure, see www.iste.co.uk/paret/nfc.zip*

Figure 7.21. *Example of an antenna shielded by a sheet of ferrite, installed on the metal casing of a mobile phone battery (Google Nexus)*

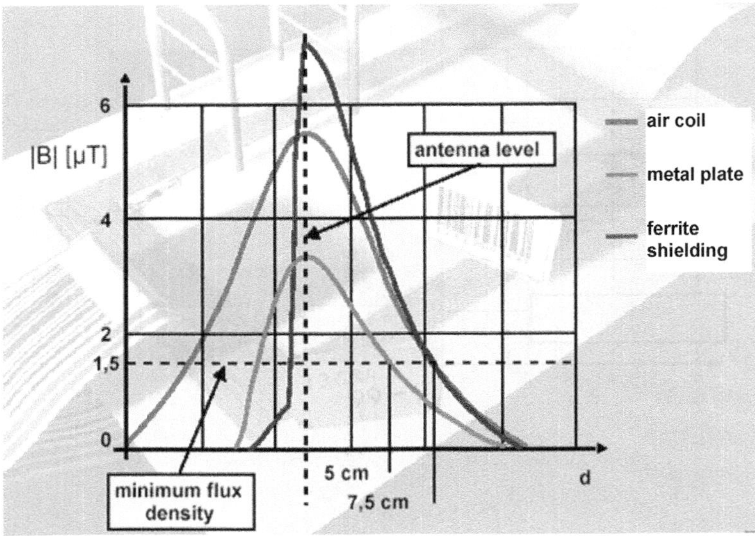

Figure 7.22. *Comparison between the different solutions shown above. For a color version of the figure, see www.iste.co.uk/paret/nfc.zip*

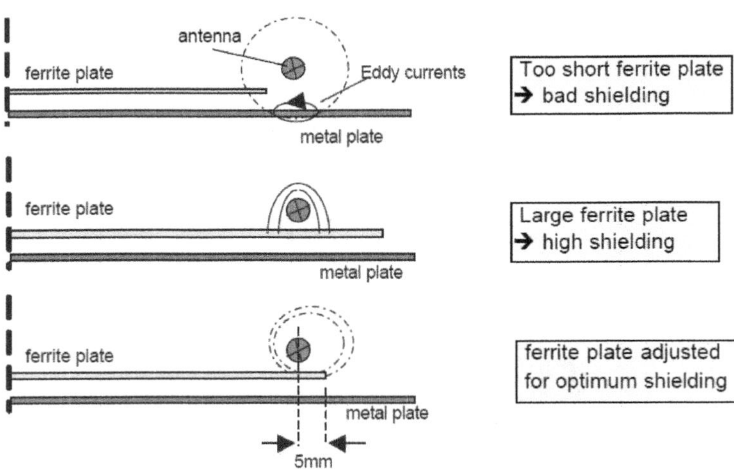

Figure 7.23. *Effects of the positions and relative distances between the metal shielding and the ferrite plate*

PART 4

Conformity and Interoperability Constraints

Introduction to Part 4

This final part of this book, devoted to the constraints on measurements and tests of NFC systems and their antennas, will lead us into a discussion of the resulting conformity and interoperability (or non-conformity and non-interoperability).

It is divided into two chapters: the first is devoted to the constraints of measurements of the values and characteristics of the various NFC-specific components (antennas, coils, etc.), and the second describes the main constraints of conformity tests to be satisfied, if we want a product to conform to a certain standard (whichever standard that may be).

8

Conformity Constraints

This chapter discusses the main methods for taking measurements and their constraints to characterize and develop the initiators and targets of an NFC system.

8.1. Conformity tests for NFC devices

The international standardization bodies (ECMA, ISO, CEN, ETSI, etc.) and the issuers of proprietary standards (EMV, NFC Forum, etc.) are in charge of producing the documentation and the tests to be passed in order to gain certification a comply with the generic norms/standards. In summary:

– the first step is to write a norm/standard and the values to be respected which we wish to include in that standard;

– next, in order to minimize the number of interpretations that the different methods and measuring tools could raise, we write another norm (or several) known as "conformity tests", describing how (the procedures to be implemented, etc.) and what to use (which devices, equipment, value calculation methods, etc.) to measure and evaluate the values indicated in the generic standard. This document merely serves to verify whether the values indicated in the fundamental standard are respected (e.g. if the designers of the product being tested have properly understood the protocol and responded correctly to it in accordance with the standard document). However, it absolutely cannot describe the performance(s) of the communication, and whether that communication is established over a distance of 3 cm or 3 m, for example, except in the case of certain sector-wide proprietary standards (EMV, NFC Forum, etc.).

8.2. *Norms of "conformity" tests for NFC IPx*

Having made these generic remarks, Table 8.1 shows the official norms and standards of conformity pertaining to the domain of application linked to NFC.

Organizations	Working group	Applications	Fundamental norms	Conformity norms	Performance norms
ISO JTC1	SC 17 WG 8	Proximity cards	ISO 14443	ISO 10376-6	–
		Vicinity cards	ISO 15693	ISO 10376-7	–
ECMA	TC 47 (TC 32 TG19)	NFC-IP1	ECMA 340	ECMA 356 RF	–
				ECMA 362 Protocol	–
		NFC-IP2	ECMA 352	–	–
ISO JTC1	SC 06	NFC-IP1	ISO 18092	ISO 22536 RF	–
				ISO 23917 Protocol	–
		NFC-IP2	ISO 21481	–	–
		Sector of application	Fundamental standards		
NFC Forum		NFC Forum	Analog	Candidate V2	–
			Digital protocol		–
CEN	TC 278	Public transport	TS 16 794 – 1		
			TS 16 794 – 2		

Table 8.1. *Official conformity norms and standards*

NOTE.–

1) The NFC conformity tests pertaining to ISO 18092 (NFC IP1) and ISO 21481 (NFC IP2) are described in the two norms ISO 22536 and 23917. For the most part, they stipulate the physical values of the radiated- and received-fields, the modulation indices, the field values, retromodulation and all the numerous timings associated with communication (bit, frame, etc.), which are included in this book, in direct relation with NFC antennas.

2) Except for the chapter on the characteristics of products using FeliCa technology, given that the physical layer in ISO 18092 is not hugely

different from, and indeed is heavily inspired by, the ISO 14443 norms, the conformity standard ISO 23917 is similar to the conformity standard of ISO 14443 – i.e. ISO 10373-6 – barring a few (not-insignificant) exceptions.

In order to be certified as "compliant", a product must be *100% compliant* with these norms (and there are many to conform to), and in order to conform to them, it must undergo numerous conformity tests (to give a rough idea, there are over a thousand such tests) and, most importantly, "*pass*" all of them, and "*fail*" none. For simplicity's sake, let us also state briefly that, in order for these tests to exist, there must be testing tools, which must be representative of real-world applications. However, as has been demonstrated throughout this book, these applications are extremely numerous, and above all are hugely varied, which somewhat complicates the task of defining test tools and the measuring procedures with numerous options in terms of hardware (lower layers) or software (higher layers) from the OSI model. Additionally, in order to carry out these tests and make them official, there have to be independent laboratories which are approved and accredited by a national body (e.g. COFRAC in France), rather than laboratories which are "more or less independent", in which no conflict of interest exists. It would be much easier to approve one's own products without outside influence, endowing them with a seal of approval and an attractive logo.

8.2.1. *ISO 23917 tests (NFC IP1 protocol)*

Let us begin with this norm, because it contains nothing exceptional which needs to be discussed in this book, merely the conventional digital protocol test.

8.2.2. *ISO 22536 tests (NFC IP1 RF interface)*

As mentioned previously, the conformity tests are derived from ISO 10373-6, essentially written to test "chip cards" or "chip card-style" objects (which essentially are chip cards, all things considered) with an array of measurements devoted to plastic chip cards in ID1 format. The inherent problem with NFC (and it is indeed inherent) is that the NFC IP1 ISO 18092 and IP2 ISO 21481 norms do not mechanically define what an NFC device is, leaving the way open for any and all form factors which designers may dream of. Obviously, it would be meaningless to try to measure all of these creations in an ISO array strictly devoted to chip-cards and, since the dawn of NFC, its limitations have been apparent. Evidently, since then, the ISO has altered its specifications creating six distinct classes (see ISO 10373-6.) For the rest, there is absolutely nothing official.

To return to the problem at hand and to the technical questions, ISO 22536, like ISO 10373-6, notably stipulates measurements of minimum values of the magnetic fields H emitted in initiator mode (remember that $H\_min = 1.5$ A/m), and a minimum value of the retromodulated voltage. Why is the emphasis placed primarily on these two parameters, but not others? The answer is very simple: NFC can only function if its physical layer is working, and thus at the very least, these two values must be satisfied. Unfortunately, this is often where the problem lies.

Let us once again return to one of our favorite examples.

Example of field radiated by a mobile phone at distance zero:

H_0_front screen side = 1 A/m rms

H_0_back battery side = 0.5 A/m rms (impact of the loading effect)

In all cases, assume we have an emitted magnetic field H weaker than 1.5 A/m. This mobile phone, therefore, will never conform to the norm and can never bear the label "ISO-compliant NFC". It is not even worth the trouble of running through the long list of conformity tests for the digital protocol. The "conformity" of that NFC device then becomes a "proprietary specification", with no guarantee of interoperability from one product to the next and/or one client to the next. In their defense, it must be acknowledged that the marketing teams of most mobile phones manufacturers are often forced to satisfy the formula "phone function – all gadgets – energy consumption – battery life – launch on the market" and that at present, the value of the field radiated for NFC is not a priority for them ... but finally, this is beginning to change. Nevertheless, this is not enough for NFC! Why is this?

Let us again look at the example of an NFC mobile phone amongst the legion of other possible NFC devices. How can we apply the conformity tests to the below NFC device on the array of tests?

– is the mobile phone:

- a flip phone,
- flip free,
- a slide phone,
- integral?

– is the antenna in the flip cover, in the slide section, or in the body of the mobile?

– does the NFC phone work with the flip cover/slide:

- open,
- closed,
- open and/or closed?

– is the antenna:

- on the side of the screen (i.e. the front side, level with the screen),
- on the side of the battery (i.e. the back)?

– is the antenna positioned:

- all over the mobile,
- at one end of the mobile,
- is this indicated?

– does the NFC function perform equally well from the top and the bottom of the mobile phone?

– is it necessary to run a host of conformity tests in all imaginable situations where the user might use his/her mobile phone?

The above gives a brief glimpse of the numerous issues and constraints which must be dealt with in order to be able to boast ISO 18092 compliance, and claim that a product has passed all conformity tests with flying colors.

Let us now turn to the characterization of the main parameters of the initiator- and target-antennas.

8.3. Electrical characterizations of the initiator antenna

Numerous parameters concerning the initiator antenna need to be measured and characterized. We shall begin by looking at the electrical values of its inductance and resistance.

8.3.1. *The antenna's inductance and resistance*

In principle, at 13.56 MHz, in order to characterize an antenna, it is necessary to establish its *equivalent serial diagram* comprising the elements L_a and R_a, all with a parasitic parallel capacitor C_a (see Figure 8.1).

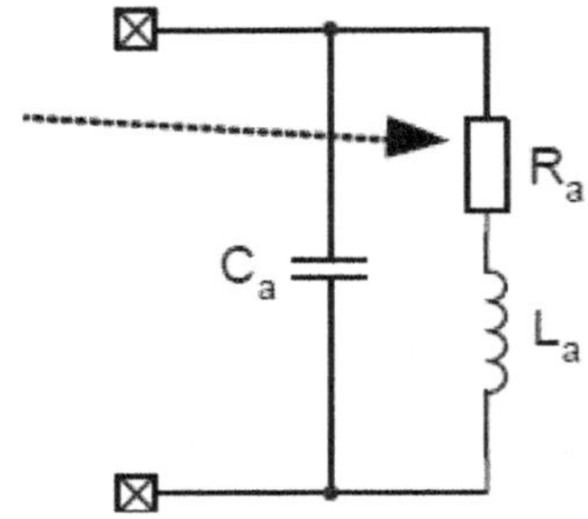

Figure 8.1. *Equivalent serial diagram*

In order to do so, we first need to examine its *physical diagram*, which is that of a coil with inductance L_a, a serial resistance of the coil R_s, and a loss resistance R_p placed parallel to the coil, all with the same parasitic parallel capacitor C_a (see Figure 8.2).

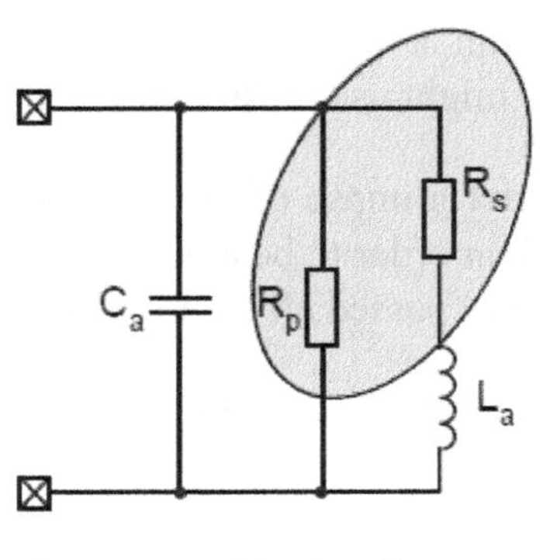

Figure 8.2. *Physical diagram*

The classic theory, recapped in an application note from NXP, indicates that with four parameters measured by a *vector network analyzer (VNA)* or a bridge or a Smith chart, it is possible to deduce the equivalent diagram. These parameters are:

– L_a: equivalent inductance at f = 1 MHz;

– R_s: equivalent resistance at f = 1 MHz, but with no skin effect;

– R_p: equivalent resistance at the specific resonance frequency;

– f_{ra}: the antenna's specific resonance frequency.

From the measured value of the frequency f_{ra}, we can immediately extract the value of the parallel capacity:

$$C_a = \frac{1}{(2 \cdot \pi \cdot f_{ra})^2 L_a}$$

The VNA measurements and Smith chart readings (Figures 8.3(a) and (b)) can be used to extract other results.

As we are now in possession of the measured values of R_s and $R_p(f_{ra})$, the equivalent value of the series resistance R_a of the antenna at the working frequency of 13.56 MHz can be calculated in order to construct the equivalent diagram on the basis of the physical one:

– the value of the parallel resistance $R_p(f_{ra})$ measured on the Smith chart enables us to calculate its value R_p@13.56 MHz:

$$R_p(13.56\ MHz) = \frac{R_p(fra)}{\sqrt{\frac{13.56}{fra}}}$$

– supposing that the value of the quality coefficient Q is high (which is very often the case), the value of R_a@13.56 MHz on the equivalent diagram is calculated by using R_p@13.56 MHz in the approximate formula:

$$R_a = R_s + \frac{(2 \cdot \pi \cdot f_{op} \cdot La)^2}{R_p(13.56\ MHz)}$$

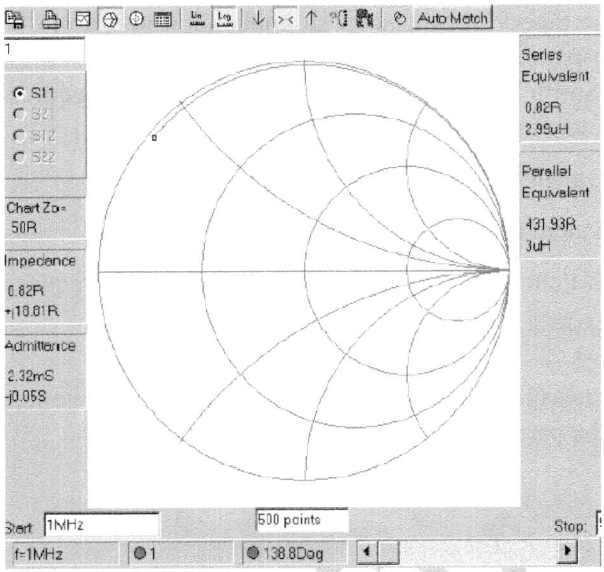

a. R_S = 0.82Ohm, L_a = 2.99 µH

Values of R_s and L_a measured at f = 1MHz

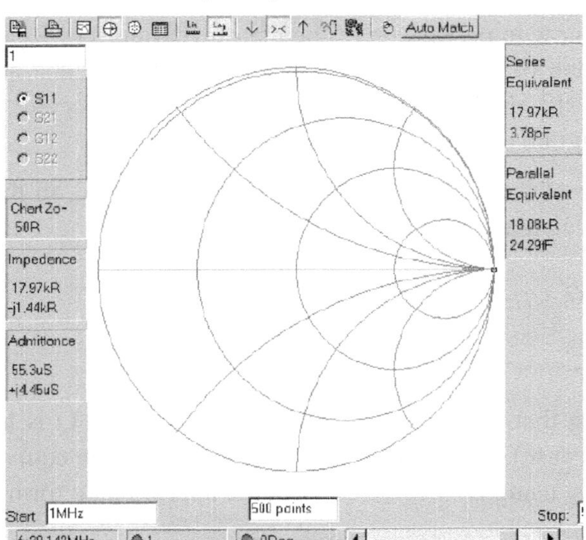

b. R_p = 18kOhm, f_{ra} = 29.14MHz

Value of R_p measured at the frequency f_{ra}

Figure 8.3. *VNA measurements and Smith chart readings*

8.3.1.1. *Skin effect of the antenna's resistance*

In actual fact, the antenna's resistance is not very easy to estimate, because the skin effect may begin to manifest itself at 13.56 MHz, thus leading to erroneous values of the quality coefficient Q. See the final section of this chapter for full details.

8.3.2. *Quality coefficient*

With a fairly accurate knowledge of the two values L_a and R_s, we can extract the theoretical value of the quality coefficient Q in the application. This estimated value of Q is, of course, the "no-load" value, measured in air, with no loading effect, which will never occur in real life, and the measurements will need to be taken in the harsh light of day (again, see the end of the chapter for complete details).

8.3.3. *Bandwidth*

Here, we see the same problem, for the same reason, but armed with the value of Q, it is easy to estimate the most likely bandwidth Bp using the classic formula Bp = fo/Q.

8.3.4. *Tuning of the antenna and its impedance matching*

We have, at great length, discussed the point of the strong constraint of the quality of impedance-matching circuit of antennas for initiators (EMC filtering and capacitance bridges) in many previous books by this author (see the footnote in the introduction to this book). Now we need to check the veracity of our ideas. There are various possible methods to do this.

8.3.4.1. *So-called substitution method, using Lissajous curves*

The method described below is a simple way to verify the tuning of the antenna's impedance-matching circuits (see Figure 8.4). It begins with the very simple step of calibrating a series of measurements using a sine wave generator at 13.56 MHz and a dual-trace oscilloscope, and then presenting the setup to be tested, by substitution in this array of measurements (inserting the values for the antenna and the impedance-matching components) and adjusting the values of the components until we obtain same values as the initial calibrated setup.

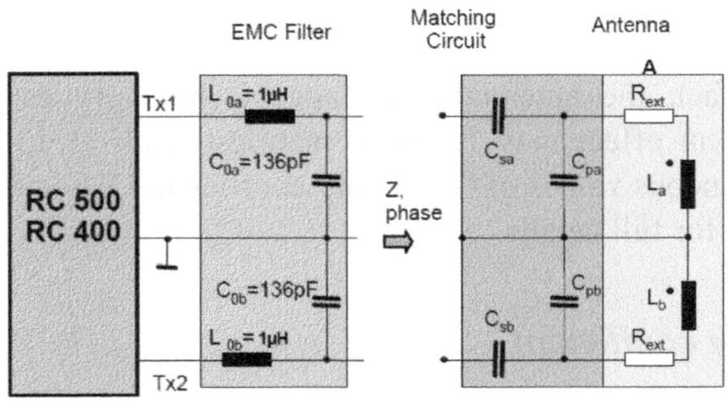

Figure 8.4. *Impedance-matching circuits of the antenna*

8.3.4.1.1. Calibration of the measurement array

The setup used for calibrating the array is presented in Figure 8.5. It is none other than a compensated RC divider bridge (called an "integro-differentiator filter"), like that which is supposed to be used as a probe for measuring the division by 10 of an oscilloscope with its input impedance.

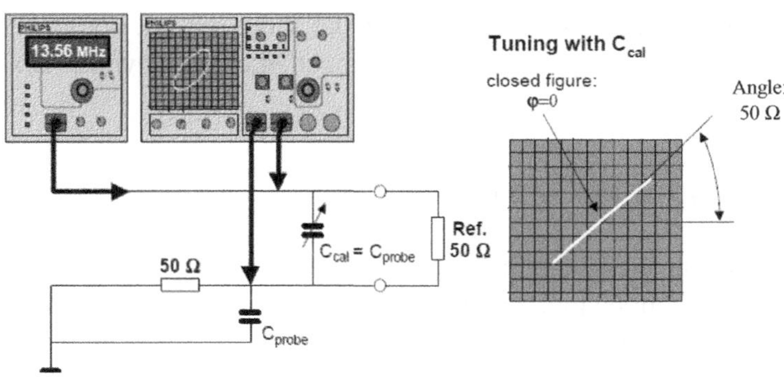

Figure 8.5. *Setup used for calibrating the array*

Suppose, for example, that we have chosen to use an integrated circuit whose output resistance, R_ic_out, is 50 ohm; in this case, the whole impedance-matching setup of the antenna also needs to have an impedance of 50 ohm.

Method of calibrating the array

To begin with, we arrange two identical resistors in a serial setup (e.g. 50 Ω as shown in the figure).

We take care to verify the calibration of the two probes of the X and Y channels of the oscilloscope, by placing the probes at the same measuring point (on the same sine-wave voltage) to obtain, when observing Lissajous curves, a straight line at 45° (this is also the classic phase of adjustment to X and Y = f(t) of the capacities of the square-wave signal probes).

On the channel which is on the side of "resistance connected to ground", we place an oscilloscope probe "number 1" (i.e. 1 MΩ), whose exact capacity C_{probe} is unknown but close to 10 pF.

On the other resistance (that which is connected directly to the input of the second channel), we experimentally determine the value of the capacity C_{cal} (essentially 10pF, for example, by twisting two isolated wires together), so that we obtain a Lissajous curve showing a 45° straight line rather than an ellipse, and consequently, the same capacity value as that of the probe.

NOTE.– When we wish to get rid of that part of the circuit specific to EMC filtering and keep only the true load of the output antenna (for instance, having chosen a load of 700 Ω, as shown in Figure 8.6), we replace the 50 Ω resistances from the previous setup (indicated in the figure for 50 Ω matching), with 700 Ω resistances. The goal is still to generate a Lissajous curve at 45° between the two signals, having taken care to compensate for the value of the capacity of the oscilloscope probe.

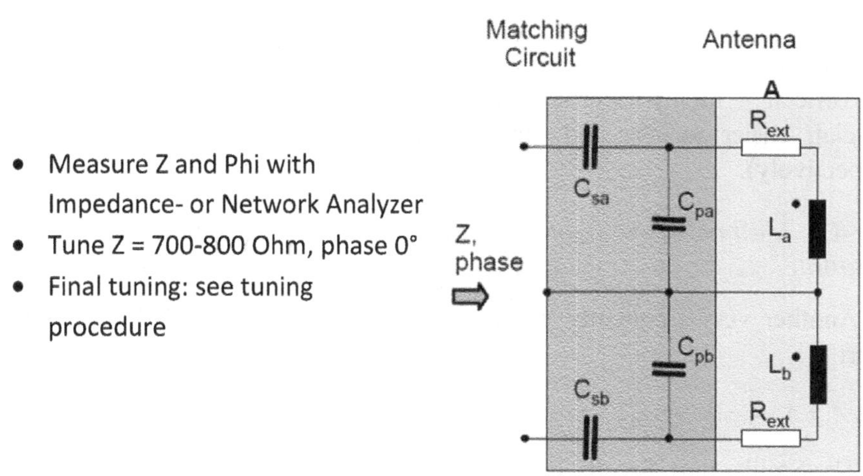

Figure 8.6. *Example of 700 Ω matching*

8.3.4.1.2. Measurement of detuning

Once the value of the capacity (C_{cal}) has been determined, we need to remove the 50 Ω resistance from above and replace it with the "global antenna circuit" whose impedance (on paper) should be Z = 50 + j0 which should yield the same result. One can always dream. Thus, we have an ellipse, rather than a straight line. (see Figure 8.7).

Thus, we need to adjust/modify the values of C_2 and C_1 so as to perfect the matching.

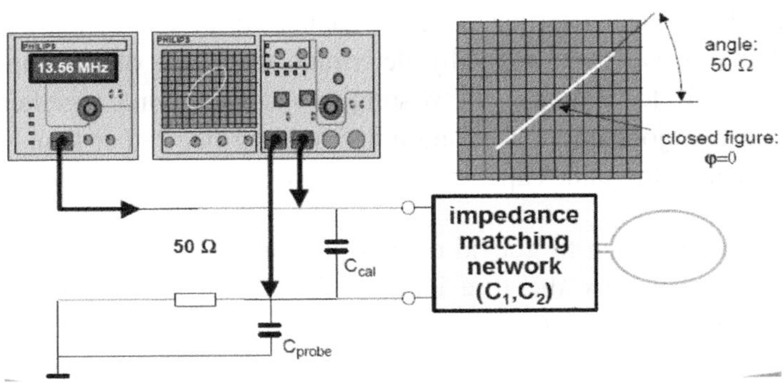

Figure 8.7. *Measuring the detuning and matching settings*

8.3.4.1.3. Interaction of the compensation values

Two possibilities arise to resolve this stage of the measurements: one official – highly rigorous but painstaking – and one which is almost official, easier to use and almost perfect (see Figures 8.8(a) and (b) respectively).

8.3.4.2. *Method of verification of tuning and adjustment of the initiator antenna*

Another very frequently used, more visual method is VNA and Smith chart.

8.3.4.2.1. Smith chart

The Smith chart directly indicates whether the setup is tunable or whether it is primarily inductive or capacitive which may be a goal in its own right, depending on the types of detuning needed for the target(s) (stacks, etc.) and their applications (see Figure 8.9).

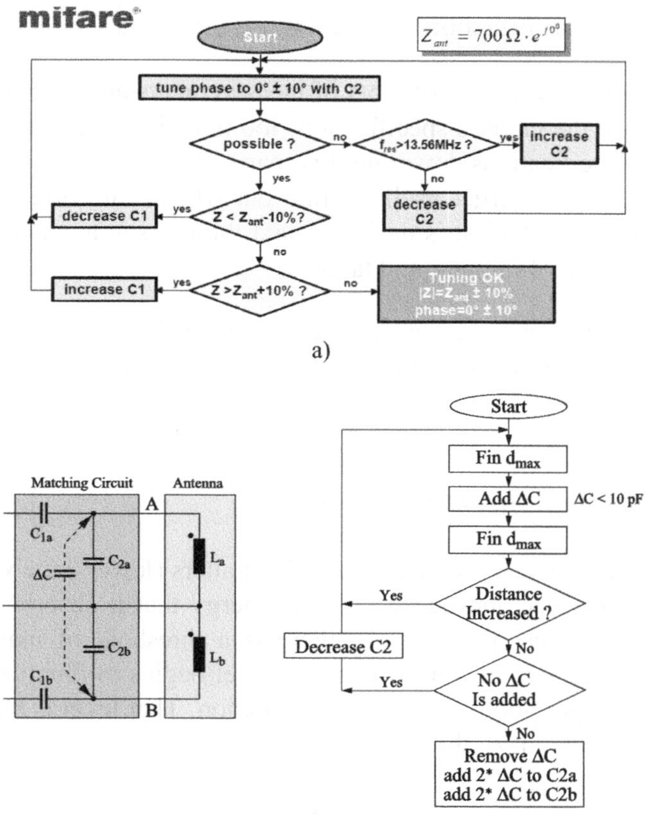

Figure 8.8. *Interaction of compensation values*

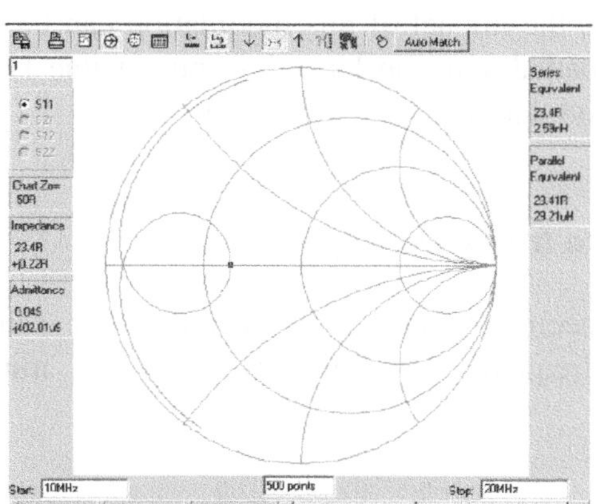

Figure 8.9. *Smith chart (strictly tunable setup)*

8.4. Method of adjustment of the target antenna

Now we come to the constraints of measurements and adjustment of the target, because the target's specifications too must be met. It is necessary to break these specifications down into two parts: the constraints relating to the performances of the target itself, to find out whether it corresponds to the specifications, for which it is designed, and the measurements pertaining to the overall system of initiator and target.

To characterize a target, it is necessary to define a certain number of parameters so as to then measure them. For example, we can cite the values of measured magnetic induction, the resonance frequency of the target, etc.

8.4.1. *Absorption threshold, H_thr*

When a target enters the magnetic field, it alters (lowers) the value of that field, by its presence and because of the energy that it absorbs in order to operate (the loading effect). Thus, there is a threshold of magnetic field H_thr or of induction, beyond which the target begins to function and gives rise to energy absorption, reducing the induction. That level is known as the "absorption threshold, H_thr".

8.4.2. *Absorption in reading mode, H_read*

Similarly, when the target is in reading mode, it modulates its load (PLM or ALM), and thus the impact of this modulation can be measured between the two states – loaded and non-loaded – and a value H_read showing the degree/index of modulation can be measured.

8.4.3. *Resonance frequency, f_res*

The target's specific resonance frequency, f_res, can be measured by determining the frequency at which we find the minimum absorption threshold H_thr.

NOTE.– As is often indicated, for very numerous reasons of application, it is not compulsory for the specific resonance frequency of the target to be the same as the carrier frequency delivered by a system's base station.

8.4.4. *Bandwidth*

By taking the above measurement and reading the values of the curve H_thr as a function of the incident frequency, we can determine the bandwidth at -3dB of the target easily, and thus the quality coefficient Q of the target (which is different to that of the initiator), using the formula Bp = f_res/Q.

8.5. Measuring methods for use with the target

In order to measure the parameters listed above, we must have a test array creating a homogeneous magnetic field, with Helmholtz coils, made up of a pair of "field-generating coils", electrically connected in a series and arranged in magnetic phase so that their orientation creates a uniform magnetic field between the coils. The intensity of that field depends only on the mechanical form of the coils (radius, number of spires, etc.) and the voltage applied to them.

Then, we place two pairs of coils on either external side of the field-generating coils. They are used to measure the impact of the presence of the target "differentially":

– the first is the pair of "reference coils", which are set up in series and in phase, and are used (as the name indicates) as a reference for the measurement during the modulation of the target. Indeed, as they are situated a fair distance away from the center of the setup, they will hardly be affected by the presence of the target, and the recorded voltage serves as a measurement reference;

– the other pair which is used for the measurement is the pair of "sensing coils". They are also linked electrically in a serial configuration and are in magnetic phase. They too are placed in a Helmholtz configuration so as to minimize the different variations of position of the targets in the measurement. The voltage induced at the terminals of these coils is proportional to the induction.

As the modulation has very little impact on the field produced by the field-generating coils, the set of the "sensing" and "reference" coils is configured as a "bridge", to obtain the output signal serving for the measurement. In the absence of the target in the field, the reference coils are used to minimize the output voltage by zero-balancing the bridge.

Once all of the above has been done, for a given voltage V applied to the field-generating coils (of known impedance and quality coefficient Q), it is possible to calculate the maximum induction, B_max, which will be available to the mechanical center of the setup and, in view of the total surface area and the coupling represented by the sensing coil, the sensitivity will be x μT/V.

With regard to targets and initiators corresponding to the ISO 18092, 21481, 14443-2 and 15693 norms, the norms in the ISO 10373-6 and -7 series describe the tools and test setups (calibration coil, reference readers, reference cards, etc.), the testing methods (electrostatic discharge – ESD – and load modulation) and methods for testing the base stations (communication volume, amplitude modulation, reception of load modulation, etc.). Readers can refer to the official texts for any further information on this subject.

8.6. Electrical measurements of the initiator antenna

8.6.1. *Measuring the magnetic field H radiated by the initiator*

It has been mentioned in this book frequently that we need to refer to magnetic field values H. There are numerous methods to measure those values but, in the context of the set of NFC applications described in this book, pending most exact measures in a certification laboratory, the author recommends a very simple experimental method to measure the magnetic field H_0 produced at the center of the initiator antenna, which is to use an oscilloscope probe (> 1 MΩ, < 14 pF) to measure the value of the voltage induced, present in an open circuit at the terminals of a coil known as the "ISO 10373-6 *reference coil*". This coil is easy to make, and its characteristics, which are easily reproducible, are given in Table 8.2.

		Units	
Shape of the spire			Rectangular
Dimensions		mm	72 × 42
Number of turns			1
Track	Material		copper
	Thickness	µm	35
	Width	µm	500
at 13.56 MHz	Inductance	nH	250
	Resistance	ohm	~ 0.4

Table 8.2. *Characteristics of the coil known as "ISO 10373-6 reference coil"*

NOTE.– This antenna is designed for a standard chip-card format, and will, of course, yield results which are slightly incorrect for other formats used in other applications, but the order of magnitude will always hold true.

In addition, the below calculation details the value of V_induced = f(field H) of the ISO reference coil. With the above inductance values, we obtain the following (see Table 8.3).

Parameters	Units	Values or equations	Comments
f	MHz	13.56	
ω	rd/s	$85.157\ 10^6$	
H_r.m.s.	A/m ave. r.m.s.	1	
i(t)		= I_max sinωt	
v(t)		= L di/dt = L I_max ω cosωt	
V_r.m.s.	V_r.m.s.	= L I_r.m.s. ω	
Φ_r.m.s.	Weber/m²	= L I_r.m.s.	total magnetic flux, averaged over the surface of the loop
e_r.m.s.	V_r.m.s.	= Φ_r.m.s. ω = B_r.m.s. S ω = µ H_r.m.s. S ω	
H_r.m.s.	A/m r.m.s.	1	applied field
U_r.m.s.	V_r.m.s.	$= (4 \times 3.14\ 10^{-7}) \times 1 \times (72 \times 42)10^{-6} \times 85.157\ 10^6$ = 323.43 mV_r.m.s.	applying a field H of 1A/m r.m.s.
U_r.m.s.	V_r.m.s.	$= \sqrt{(323.43^2\ mV + 0.4^2\ mV)}$ = 323.43 mV_r.m.s.	
→ this represents a sensitivity of 900 mV_pp in an open circuit per A/m r.m.s.			

Table 8.3. *Value of V_induced = f(field H) of the ISO reference coil*

NOTE.– This reference coil can also serve to measure the loading effects to which a target is subject.

8.6.2. *Measurement of the quality coefficient – Q*

Many times in this book, we have estimated the theoretical values of the quality coefficient – Q – but, whatever we say or do, the most interesting thing is the true value, which we measure in the real application, including the loading effects, of the environment (the casing, the battery, etc.), which are often difficult to pre-estimate. In addition, the value of Q has a twofold influence:

– on the shape of the cutoffs and rises of the magnetic fields, particularly when the bit stream is paused;

– on the value of the bandwidth, and thus on the achievable bit rate.

Hence, before carrying out this measurement, this means we need to know a few details and also gain a certain amount of theoretical feedback.

8.6.2.1. *Q and the cutoff of the field*

Let us now examine the parameters influencing the cutoff of the carrier frequency during the "pauses" in ASK 100%, and the variations of the modulation in ASK 10%. In order to do so, we shall examine the pulsed response of the antenna circuit.

During an ASK 100% or m% amplitude shift, when a voltage-step-type pulsed transition E is applied to a serial R, L, C antenna circuit, called a "damped resonant circuit", the equation of the current running through the circuit during the transition, in the form of the Laplace–Carson transform, is of the well-known form:

$$I = \frac{E}{L} \times \frac{p}{\left(\omega 0^2 + 2\Delta p + p^2\right)} = \frac{E}{L} \times \frac{p}{\omega 0^2 \left(1 + \frac{2\Delta p}{\omega 0^2} + \frac{p^2}{\omega 0^2}\right)}$$

We state $2\,\Delta = R/L$, and set:

f0 = carrier frequency

$R < \sqrt{(4L/C)}$

$\omega 0 = 1/\sqrt{LC} \Longrightarrow L\,C\;\omega 0^2 = 1$

$$\omega 0 = 2\ \pi\ f0$$

$$T0 = 1/f0$$

After inverse transformation, the resulting signal causes a variation in voltage (an increase or decrease depending on the polarity of the variation), exponential of the amplitude of the oscillations of period T0. Apart from a few second-order approximations, when the excitation is stopped or reapplied, the mathematical expression of the variations in the voltage at the terminals of the inductance of the oscillating circuit is presented in the following form:

$$v\ (t) = Vmax \times \cos \omega\ t \times [e^{(-\ \Delta\ t)}]$$

where the quantity "$[e^{(-\ \Delta T)}]$" is called the "logarithmic decrement".

This variation, then, is a cosine wave with an exponential envelope whose time constant is equal to $q = 1/\Delta\ = 2L/R$. Thus, the equation can be written as follows:

$$v(t) = Vmax \times \cos \omega\ t \times [e^{(-\ t/q)}]$$

Figure 8.10 shows an example.

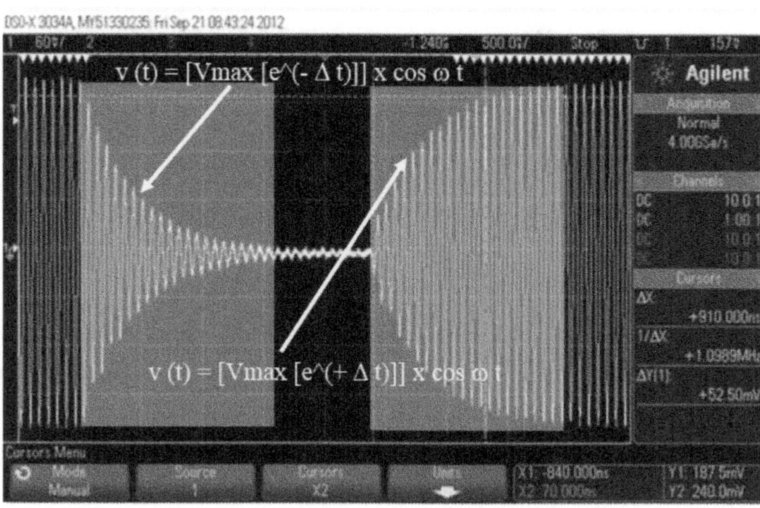

Figure 8.10. *Shape of voltage at the terminals of the antenna inductance*

Knowing that for the initiator antenna, $Q1 = L1s\ \omega0/R1s$, we can also write that:

$$Q1 = 2L\,(2\,\pi \times f0)/2R = 2L\,(\pi \times f0)/R$$

However,

$$q = 1/\Delta\ = 2L/R$$

$$q = Q1/(\pi \times\ f0)$$

$$= (Q1/\pi) \times T0$$

Consider the relation between q and Q1:

$$\frac{q}{T0} = \frac{Q1}{\pi}$$

Let us calculate the values of the voltages at times t1= q and t3 = 3 q:

– for t1 = q:

$$v(q)\ = Vmax \times e^{-1}$$

so $$v(q)\ = Vmax/2.718$$

so $$v(q)\ = 36\%\ \text{of}\ Vmax$$

– for t3 = 3 q:

$$v(3\ q) = Vmax \times e^{-3}$$

so $$v(3\ q) = Vmax/20$$

so $$v(3\ q) = 5\%\ \text{of}\ Vmax$$

This means that, at t3 = 3 q, the amplitude of the signal is only 5% of its initial maximum amplitude. If the initiator works with ASK 100% amplitude shift keying:

that time t3 = T_cutoff at 3 q

$$= 3 \times (1/a)$$

$$= 3 \times (2L/R)$$

is that which is necessary for the carrier frequency to be physically (nearly, up to 5%) cut off.

At the setup's working frequency (f0, T0 = 1/f0, ω0 = 2 π f0), we can also find the numerical value of T_cutoff_x% as a function of the quality factor of the circuit at that frequency, Q1 = L ω0/R, which gives the following:

$$\text{As } Q1 = 2L\,(2\,\pi \times f0)/2R = 2L\,(\pi \times f0)/R$$

$$T\_cutoff\_36\% = 1\,q = (Q1/\pi\ f0) = (Q1/\pi) \times T0$$

$$T\_cutoff\_5\% = 3\,q = 3 \times (Q1/\pi\ f0) = 3\,(Q1/\pi) \times T0$$

8.6.2.2. *Decrease of the field according to the ISO norm*

Figure 8.11 shows the example of the NFC IP1 and IP2 norms (ISO 18092 and 21481), which conform to the ISO 14443 A norm.

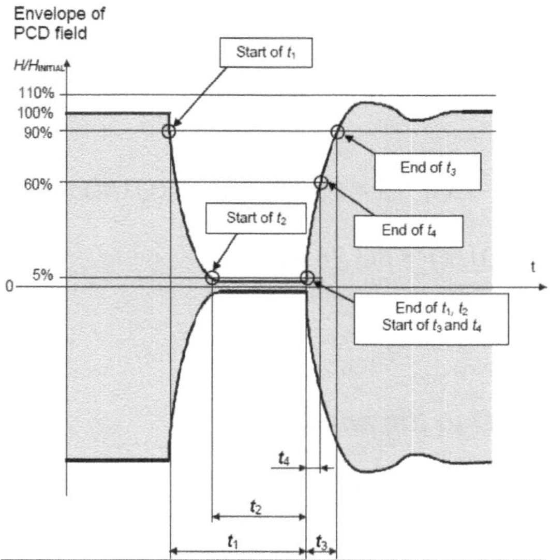

Parameter	Min	Max
t_1	28/*fc*	40,5/*fc*
t_2	7/*fc*	t_1
t_3	1,5 × t_4	16/*fc*
t_4	0	6/*fc*

Figure 8.11. *Decrease in field according to the ISO norm*

$(t2 - t1)max = (40.5\text{-}7)/fc \sim 42/fc$

$= 73.7\ ns \times 42 = 3\ \mu s$

$(t2 - t1)min = (28\text{–}7)/fc = 21/fc$

$= 73.7\ ns \times 21 = 1.5\ \mu s$

This represents an average time of 2.2 μs, which is the equivalent of ~ 31 periods of the carrier frequency, hence the value of Q for NFC ISO 18092 and ISO 14443.

$T\_cutoff\_5\% = 3\ q = 3 \times (Q1/\pi\ f0) = 3\ (Q1/\pi) \times T0$

$3000ns = 3\ (Q1/\pi) \times 73.74\ ns$

$Q1 = 42.58$

$T\_cutoff\_5\% = 3\ q = 3 \times (Q1/\pi\ f0) = 3\ (Q1/\pi) \times T0$

$(average)\ 2200ns = 3\ (Q1/\pi) \times 73.74\ ns$

$Q1 = 31.22$

$T\_cutoff\_5\% = 3\ q = 3 \times (Q1/\pi\ f0) = 3\ (Q1/\pi) \times T0$

$1500\ ns = 3\ (Q1/\pi) \times 73.74\ ns$

$Q1 = 21.29$

8.6.3. *Measuring Q in the application*

Let us take an example at 13.56 MHz with a quality factor Q1 of 31.4, which is $Q1/\pi = 10$, and for $3(Q1/\pi) = 30$:

Carrier frequency f0	13.56 MHz
T0 = 1/f0	73.75 ns

where Q1 = 31.4:

– i.e. $(Q1/\pi) = 10$:

T_cutoff_36% 0.737 μs, which is around 10 alternances of the carrier frequency

– i.e. $3(Q1/\pi) = 30$:

T_cutoff_5% 2.2 μs, which is around 30 alternances of the carrier frequency

It is noteworthy that the value of T_cutoff_5% at 13.56 MHz in this example complies with ISO 14443 Part 2 A and NFC ISO 18092 (Figure 8.12).

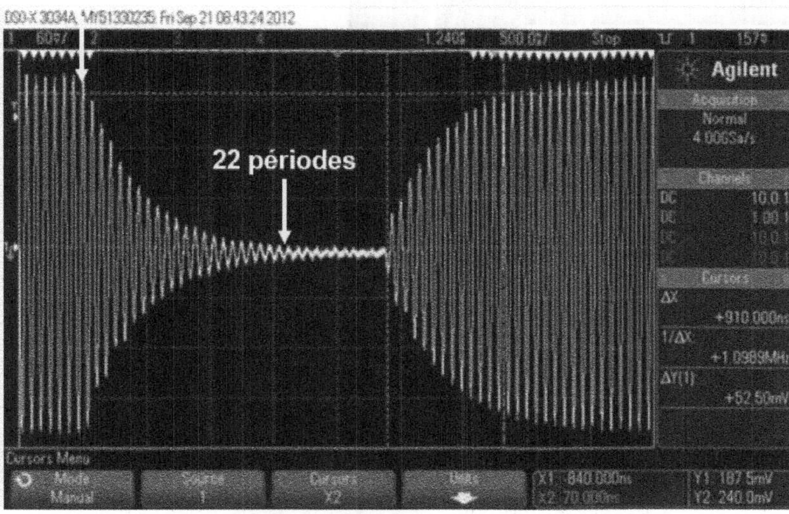

Figure 8.12. *Example of conformity with the ISO 14443 -2 A and NFC ISO 18092 norms*

One practical consideration: in a mobile environment where a screen and/or a ferrite sheet is necessary, a quality factor higher than 35 is highly unlikely. When a factor Q of 35 cannot be achieved in a given environment, it is nonetheless advisable to try to keep it at least equal to 15, in order to obtain good performances.

8.6.4. *Measuring the bandwidth in the application*

The simplest, relatively-reliable technique is to measure the bandwidth of the no-load initiator antenna circuit in the real application by means of the amplitude spectrum of the field radiated into the air (using the *peak hold function* of a VNA) with a small loop antenna (20 mm in diameter, placed a reasonable distance away from the antenna so as not to detune it), and simply ensure the measurement also in the presence of a target in order to gain some idea of the loading effect (see Figures 8.13(a) and (b)).

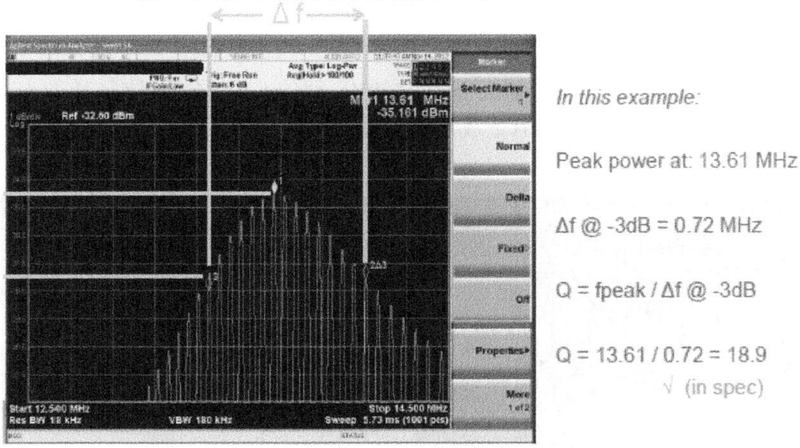

a)

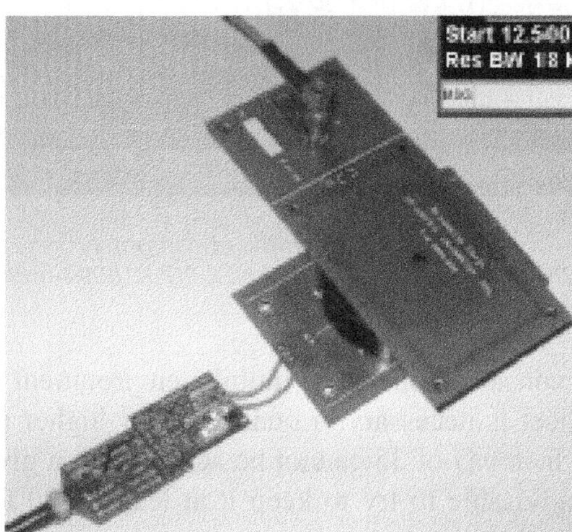

b)

Figure 8.13. *Measurement of the bandwidth in the application (document provided by Innovision)*

8.7. Method for adjustment of whole systems

In order to ensure the setup works correctly, it is necessary to carry out new series of measurements specific to the systems under development.

8.7.1. *Measurements needing to be performed*

We now come to the final set of constraints in this chapter. They pertain mainly to the overall measurements of fields and/or inductions available, which need to be carried out on the basis of the distances and the values of the coupling coefficients "k", which is the index of coupling between the initiator and target antennas.

8.7.2. *Order in which the measurements must be taken*

Is there a particular order in which these measurements need to be carried out? The answer is: yes.

We must work on the basis of the principle that the values of the inductances, the quality coefficients Q, the coupling coefficients, etc., of the different components in play are already known (calculated or measured), and that only the values of the system remain to be found.

8.7.3. *Energy domains, zero lines and safe operating areas*

Once the above measurements have been made, if the designer has not run good simulations previously, he/she may well be obliged to carry out a series of lengthy modifications to render the system reliable and reproductive. In this case, it becomes necessary to measure the Safe Operating Areas – i.e. the "energy domains" in which it can be guaranteed (by signature, of course) that the target, if it is batteryless, will actually be powered remotely. Next, depending on the degree of sensitivity of the base station demodulator, the equipotential lines may need to be measured to determine whether or not there are zero lines (inoperability due to lack of remote power).

Having reached this stage, designers can (almost) rest easy, knowing that they are very familiar with their system.

8.8. Measuring tools

There are numerous hard- and soft-development support tools on the market, to carry out all these series of measurements either manually or automatically. The best known such tools are those made, for example, by Micropross, Keolabs, Comprion, etc. (see Figures 8.14(a) and (b)).

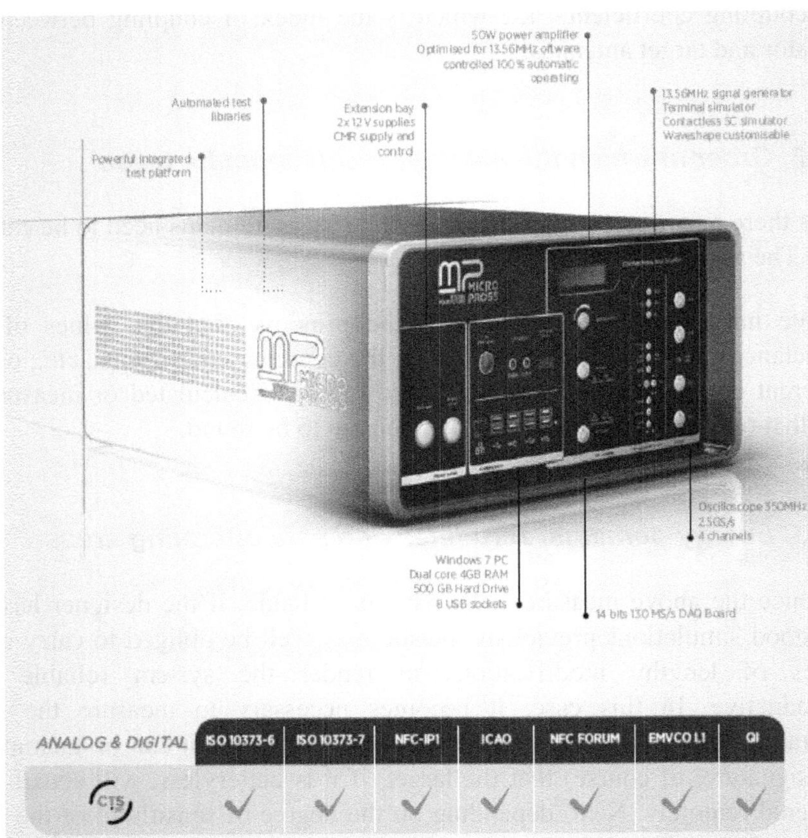

a)

b)

Figure 8.14. *Measurement tools (documents provided by Micropross and Keolabs)*

9

Interoperability Constraints

This final chapter on the constraints of interoperability of NFC systems and elements is, simply the logical progression and the consequences of the previous chapter, focusing on the constraints of conformity to proprietary standards (EMV, NFC Forum, CEN, etc.) which are partially different or adapted from the ISO's NFC norms.

Even while remaining scrupulously compliant with norms or standards, some claim to be better than others. These claims are not entirely true nor entirely false, because this attitude corresponds to, or implies, the notion of "performance", rather than strict "conformity", which many people tend to confuse. In addition, it is easy to claim to be the best when we are measuring our own performances in a particular field of application: our own. This is why everyone is always the best… in their own way. Thus the claim often means nothing beyond this. It is also for this reason that certain generic norms and conformity norms are also accompanied by "performance test" norms (such has been the case for many years in UHF RFID), which complement the conformity norms, providing us with a shared normative framework, and measurements by which to estimate the performances of products. This enables us to speak intelligently about the fine nuances between "conformities", "performances", and between "interoperability" and "interchangeability".

9.1. Norms and interoperability

It is a significant problem which we have begun to outline, looking at the main issues described above. So as to remain in the realm of concrete certainty, we shall now give an everyday example of interoperability.

What has always constituted, and continues to constitute, the strong point and success of contact chip cards ISO 7816 and contactless chip cards ISO 14443 is, incontestably, their respective interoperabilities, which can be summed up as:

– "all readers can read a particular card";

– "all cards can be read by a particular reader".

It must not be forgotten that this is due only to the fierce determination of the members of AFNOR and ISO, who established the ISO 7816-x family of norms to ensure the interoperability of contact cards and then their younger sibling, the contactless chip card (the ISO 14443 family of norms). This seems so simple that it is often all too easy to forget how many hours of debates it took to achieve in the normative and standardization committees. Ultimately, though, all of that is history; it is done, the technology exists, and it works.

In spite of the enormous diversity and overlap of the applications that NFC supports, and the very different participants behind it, in order to have the same level of success, NFC needs to reach the same level of interoperability to ensure survival over the long term; firstly for itself, and secondly for the applicational overlaps it aims to provide with applications conforming to the norms cited above. This leads directly to a hard point in the story.

9.2. Problems of the tests; NCF ISO vs EMV vs NFC Forum etc.

Besides the conventional conformity tests on the structures, protocols, and timings, the ISO's various conformity tests on NFC stem from numerous problems on the ground. Indeed, the ISO's NFC and contactless chip card norms are generic, as opposed to being specific to particular fields of application, and are essentially based on physical fundaments – notably measurements of the values of magnetic fields H. For its part, the EMV contactless proprietary standard designed for chip cards "only" stems from ISO 14443, and is adapted to the particular applications of the banking and payment sectors disregards the values linked to physics but includes in its specifications certain pragmatic values necessary for its application in the field – i.e. working distances and volumes – all being connected to simple measurements of voltage. The same is true of the NFC Forum, but with different values. Unfortunately, there is a strong chance that by 2020 payment and other applications in transport, etc., will be conducted with a mobile NFC device (e.g. a phone) rather than with a simple chip card, so we will need to resolve the problems of the values of the H_min fields, the retromodulation voltages, the form factors of the antennas, the couplings (depending on the types of NFC devices), classes of antennas, loading effects, etc.

9.2.1. *EMV conformity tests*

As mentioned above, the EMV contactless proprietary standard applies a philosophy of measurements radically different to that of the ISO because its specifications and measuring conditions indicate dimensions/volumes/distances of function to be respected, as opposed to physical values of magnetic fields.

Indeed, the official array of EMV conformity tests (see Figure 9.1) requires us to adjust or calibrate values of voltages to "x" AC volts in a specific electronic test setup (the EMV_PCD), when we do not entirely know the value of the magnetic field H generated at that time. It requires it to be able to receive on a specific EMV_PICC, arranged at particular points in space, voltages of "y" DC volts, which should come down to the same thing at some point but the question remains: "where is that point?" This is certainly a good question, but not one that is easy to answer. However, we shall attempt to explain it in the text and calculations shown below.

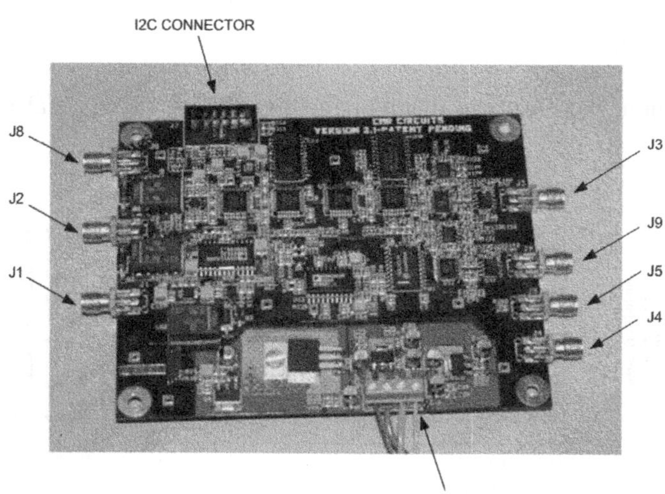

Figure 9.1. *EMV conformity test array*

Thus, the following is a verbatim reproduction of section 2.2.1 EMV – TEST PCD, from the external specification text EMVCo 2.1.

9.2.1.1. *EMV – TEST PCD*

The EMVCo document states: "The EMV – TEST PCD has a circular antenna of about 7cm, which is in the small range of antenna sizes encountered in EMV terminals. The circular antenna creates a symmetric field distribution from the z-axis, which simplifies measurements. When fed with 600 mW into its 50 Ω input impedance at resonance, the EMV – TEST PCD provides *a magnetic field which is representative of most PCDs*".

Most disconcerting in this sentence are the words "representative of most", and, to supplement the table:

"The quality of the transmitter of a PICC is measured on the EMV – TEST PCD.

EXAMPLE.–

A PCD must provide a certain level of power to a PICC.

The power delivered by the PCD is measured on the EMV – TEST PICC.

The value of the power level measured on the EMV – TEST PICC must fall within range Rtx power".

In addition, the electrical diagram for the *EMV – TEST PCD* is shown in Figure 9.2.

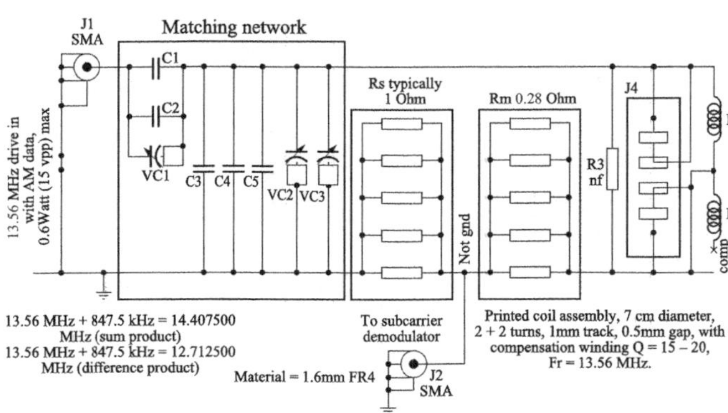

Figure 9.2. *Electrical diagram of the EMV array – TEST PCD*

The nomenclature in this diagram is shown in Figure 9.3.

Quantity	Reference	Part	Description
2	J1, J2	SMA	SMA Connector, end-launch. Johnson 1420711821. RS 363-4719
2	J4	Test point	4 holes. 2.54mm pad. No component fitted
2	C1, C4	33pF	500V, 1%. ATC100B330FW. RS 397-8170
1	C3	100pF	500V, 1%. ATC100B101FW. RS 397-8243
2	VC1, VC2	1-10pF	Johanson 250V Air trimmer type 1501, 10 turn, Q > 4000@195MHz
	VC3	Not fitted	Option for extra trimmer if required
1	L1	< 1µH	4 turn printed spiral, comp. arm. gnd rings
4	Rs1, Rs2 Rs3, Rs4	4R7	PRC 111, 1%, 0.5W, 2010 pad ref: Phycomp 23276164708
1	Rs5	6R8	PRC 111, 1%, 0.5W 2010 pad ref: Phycomp 23276166808
5	Rm1, Rm2 Rm3, Rm4	1R5	PRC 111, 1%, 0.5W 2010 pad ref: Phycomp 23276161508
	Rm5	Not fitted	2010 pad
	C2, C5	Not fitted	ATC100B pad
	R3	Not fitted	0805 pad

Figure 9.3. *Nomenclature in the EMV diagram – TEST PCD*

The EMV diagram mentions the technology used to construct the antenna and its immediate environment (see Figure 9.4) but does not indicate its value L or its serial resistance value and loss value.

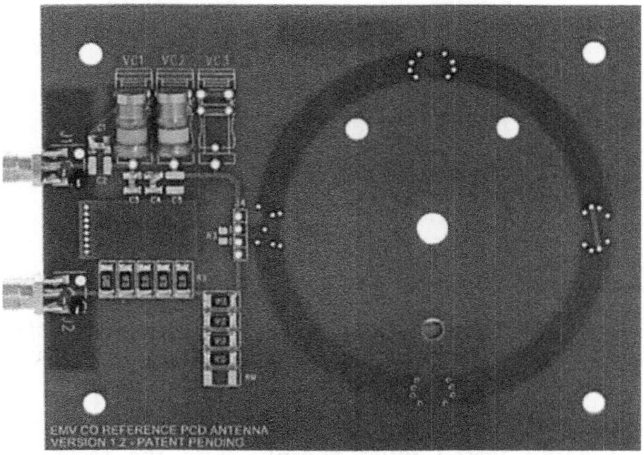

Figure 9.4. *Antenna of the EMV array – TEST PCD*

Table 9.1 Summarizes all of the above:

		Units	
Shape of the spire			Circular
Diameter		mm	70
Number of turns			2 + 2 compensation loops
Track	Material		Copper
	Thickness	µm	35
	Width	mm	1
	Gap	mm	0.5
Quality coefficient Q			15 – 20
External serial resistance		Ω	(1 + 0.28)
At 13.56MHz	Inductance	nH	
	Resistance	ohm	

Table 9.1. *Characteristics of the antenna of the EMV array – TEST PCD*

Knowing that:

– the *landing plane* (the upper part of the plastic plate) where we need to place the PICC (the target) is situated at 1.5 cm from the plane of the initiator PCD antenna (see Figure 9.5);

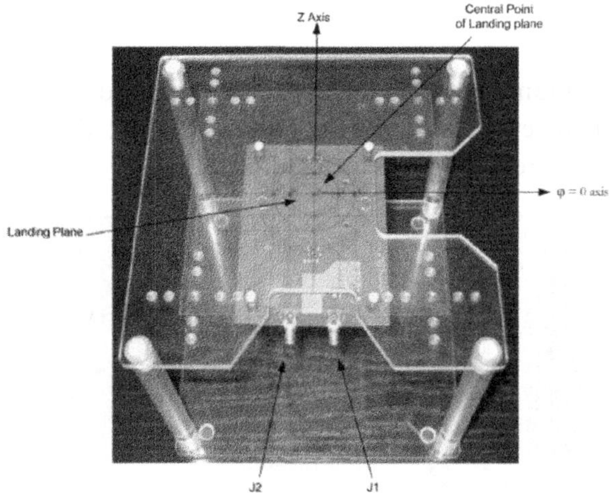

Figure 9.5. *"Landing plane" (upper part of the plastic plate of the physical plane of the PCD antenna)*

– the target being tested must comply with the tests required in the *operating volume* shown below (Figure 9.6).

It is up to us to find our way.

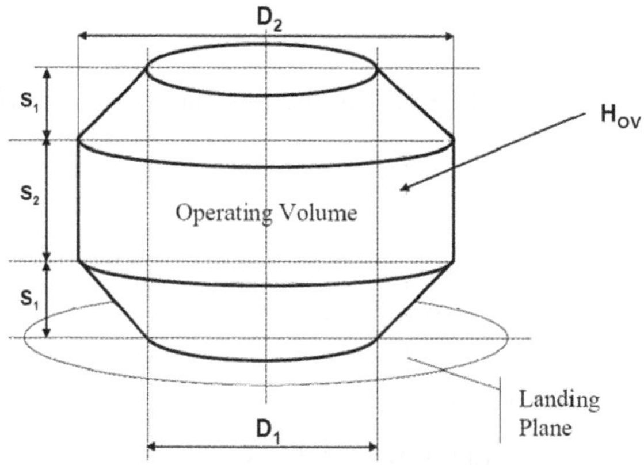

Topic	Parameter	Value	Units
Operating Volume	D_1	3	cm
	D_2	5	cm
	S_1	1	cm
	S_2	2	cm

Figure 9.6. *"Operating volume"*

9.2.1.2. *Determination of the elements*

In light of the values given in the nomenclature of the components and in the figures above, we can deduce/calculate a number of additional values.

9.2.1.2.1. Matching network

C1 + VC1 = 33 + 5 = 38 pF VC1 set to half its value

C3 + C4 + VC2= 100 + 33 + 5 = 138 pF VC2 set to half its value

9.2.1.2.2. Antenna (values calculated from the figures)

R_ant_dc = ρ l/s

$$= 1.8\ 10^{-8} \times (2) \times (2\ \pi \times 3.5\ 10^{-2})/(1\ 10^{-3} \times 35\ 10^{-6})$$

$$= 0.22\ \Omega$$

$$R\_ext\_Q = (1 + 0.28)\ \Omega$$

$$= 1.28\ \Omega$$

$$Rs\ total\_dc = R1s\ (1 + 0.28\ \Omega)\ \text{(on the diagram)} + R\_coil\ (0.22\ \Omega)$$

$$= 1.5\ \Omega$$

$$L\ calculated = 0.590\ \mu H$$

which is $L\omega = 50.24\ \Omega$ at 13.56 MHz.

9.2.1.2.3. Quality coefficient Q

EMVCo indicates a value of the quality coefficient Q between 15 and 20. Supposing we have an average Q = 17.5, this means that the equivalent serial resistance Ra of the antenna must be equal to:

$$Ra\ equiv = La\ \omega/Q = 50.24/17.5 = 2.87\ \Omega$$

Thus, there is a difference (2.87 – 1.5) = 1.37 Ω somewhere – see later on.

9.2.1.2.4. Theory

Input current for the setup

With the EMVCo data of "600 mW r.m.s. in R_in of the setup = 50 Ω", we are able to calculate:

$$P = U^2/R\_in$$

$$U^2 = P \times R\_in$$

$$= 0.6 \times 50$$

$$U = 5.47V\_r.m.s.$$

$$U = R\_in\ \ I$$

$$I = U/R\_in$$

$$= 5.47/50$$

I_r.m.s. = 109.5 mA_r.m.s.

The current in the antenna L (for further detail, see [PAR 16]) is:

$$I\_L\_r.m.s. = \frac{I\_r.m.s.}{\left|\frac{1}{\sqrt{(R\_inR1s)}} + j\frac{1}{Q1}\left(1 - \frac{1}{\sqrt{(R\_in/R1s)}}\right)\right|}$$

NOTE.– In reality, the imaginary term in “j” is very small (Q1 is often large, so the value of 1/Q1 is small, and the value of the parenthesis is small), and often I_L can be approximated as:

$$I\_L\_r.m.s. \sim= I\_r.m.s. \times \sqrt{(R\_in/R1s)}$$

$$= 109.5 \times \sqrt{(50/2.87)}$$

$$= 456.7mA\_r.m.s.$$

In principle, if the impedance matching between generators is realized, all the watt power provided by the EMVCo generator (600 mW) is transferred and dissipated into the only resistive parts of the circuit, with all the other components being reactive, then:

$$P = Ra\ I\_ant^2 = 600mW = 0.6W$$

$$= 2.87\ I\_ant^2$$

Thus:

$$I\_ant = \sqrt{(0.6/2.87)}$$

$= 0.457A$ r.m.s.... – fortunately the same value as above.

Magnetic field in H_0

The theoretical value (of the modulus of the vector) of the magnetic field (which is not entirely homogeneous across the whole surface of the antenna loop) at the mechanical center of the antenna H_0 must therefore be (with r = 3.5 cm):

$$H\_0 = NI/2\ r$$

$$= (2 \times 0.457)/(2 \times 0.035)$$

$$H\_0 = 13.05\ \text{A/m r.m.s.}$$

Magnetic field in H_0.0.0 at 1.5cm from the antenna (at point (0,0,0) on the EMV array)

Knowing that the point (0,0,0) on the landing plane of the EMV array is situated on the axis, 1.5 cm above the antenna, then given the Biot–Savart law, the field H in (0,0,0) is theoretically:

at 1.5 cm $\rightarrow$ d/r = 1.5/3.5 = 0.43

$$H = 0.775\ H\_0$$

$$H\_0.0.0 = 10.12\ \text{A/m r.m.s.}$$

Calculating the magnetic flux

In the plane of the antenna, the whole of the magnetic flux produced by the effective current I_r.m.s. circulating in the inductor of the antenna is equal to:

$$\Phi\_\text{r.m.s.\_ant} = L\ I\_\text{r.m.s.}$$

$$= 0.590\ 10^{-6} \times 457\ 10^{-3}$$

$$= 269.6\ 10^{-9}\ \text{Wb}$$

and thus, also:

$$V\_L = L\ I\_\text{r.m.s.}\ \omega = 269.6\ 10^{-9} \times 85.157\ 10^{6} = 22.94\ V\_\text{r.m.s.}$$

9.3. In practice: a few simple examples of measurements

As stated, when the EMV test array is adjusted to its nominal operating values, it is recommended to apply and calibrate the voltage applied to the EMV_PCD to "x" AC volts (on jumper 1 – J1) and, at a particular point on the EMV_PICC, receive a DC voltage of "y" V, even when we do not have any idea of the value of magnetic field generated at that time. Unfortunately, in order to design antennas for initiators and targets intelligently, it is highly practical to know the values of the fields involved. Let us give a few examples.

9.3.1. *Example 1*

Firstly at 13.56 MHz, at the input to the array EMV_PCD – the initiator – (on the jumper J1 of the official EMVCo_PDC setup), we apply a voltage of 15.5 V_pp (which represents 5.48 V_r.m.s.), and hence an input current for the EMV PCD I_r.m.s. = 5.48/50 Ω = 109.6 mA, or a power of P = 600 mW.

Then, 2cm away from the landing plane (in reality 3.5 cm away from the antenna of EMV_PCD, because that antenna is 1.5 cm above the plane of the antenna), we must position an EMC_PICC "target" whose loading effect is adjusted to a value such that we obtain and measure a voltage of 5.53 V_dc at a particular point on the EMV_PICC.

Finally, if, for curiosity's sake, we wish to gain an idea of the magnetic field that is really generated with no load in the case of the EMVCo test tool, then we withdraw on the EMV_PICC from the volume and measure the field H at the same points in space – e.g. using the voltage present at the terminals of a "calibration coil" or "reference coil" (such as that used by the ISO).

In summary, the measuring process is as follows:

1) the EMV_PCD is first adjusted to emit a "nominal field" (i.e. on the reference EMV_PICC recover 5.53V_dc at 2 cm from the landing plane);

2) the reference EMV_PICC is then placed on the EMV_PCD array at a given point in the operating volume;

3) the generator of the array is then tuned to obtain the value of the voltage Vpp min (according to the specification EMV CL 2.1);

4) the reference EMV_PICC is then deported to the PCD ISO array (in the position of the DUT – Device Under Test), which is then adjusted to emit a field equivalent to that measured at 2) at the output of the PICC.

The process is then repeated for different points in the operating volume.

An example of the results is given in Table 9.2 (these values depend heavily on the load of the target).

			The official barrel, 4 cm in height				
Point on the graph			1	2	3	4	5
Official EMV position			0,0,0	1,0,0	2,0,0	3,0,0	4,0,0
d = true distance between antennas	cm	0	1.5	2.5	3.5	4.5	5.5
H = r.m.s. magnetic field	A/m	H_x.x.x	4.9	4.5	3.5	2.6	1.75
Associated theoretical calculations							
d/r			0.43	0.71	1		1.57
H/H_0 (see formula above)			0.775	0.543	0.354		0.155
H_0... which must be a single value	A/m		6.32	8.28	9.88		11.29

Table 9.2. *Example of results*

A few interesting comments can be made about this table. Normally, one would think that across the last row, whatever the column in the table, the value of the magnetic field H_0 should remain constant. However, surprisingly these values increase progressively, which means that the effect of the couplings (distances) and the loads (consumptions) of the target – the two well-known loading effects – with ever-decreasing distances on the magnetic field produced by the initiator at H_0, and progressively its value tends toward the theoretical value calculated in air without the influence of the target.

NOTE.– Independent of these measurements, for purely informative purposes, in order to gain an idea of the magnetic field that is truly applicable in the EMVCo test described above (15.5 V_pp, i.e. 5.48 V_r.m.s. on jumper 1 J1 situated at the input of the EMV_PDC, so P = 600 mW on 50 Ω at 13.56 MHz), we withdraw the EMV_PICC from the measuring volume, and measure the "no-load" voltage V_pp at the same point in space, at the terminals of a "calibration/reference coil" such as those used by the

ISO (that which gives us the well-known value of 900 mV_pp at 1A/m rms), which gives us the following, for example:

– placed directly against the surface of the antenna of the EMV_PCD:

17.7 V_pp, so H = ~19.67 A/m

– placed in position 0.0.0 (i.e. at 1.5 cm from the EMV_PCD):

9.63 V_pp, so H = ~10.7 A/m

If, in the same way as in ISO mode, we now estimate the values of the magnetic fields H_d for the EMVCo antenna, thanks to the relation presented above, in the appendix of Chapter 1, we find (see Table 9.3):

$$H(a, r) = \frac{1}{\left[\left(1+a^2\right)^{3/2}\right]} \times H(0, r)$$

r = radius of the EMVCo antenna	Distance between the antenna and landing plane	Position in the barrel	d = true distance between the antenna and the card	a = d/r	H(0)	H(d)
Cm	Cm	–	cm		A/m	A/m
		–	0	0	19	1 × 19 = 19
3.5	1.5	0	1.5	0.42		0.73 × 19 = 13.8
		1	2.5	0.7143		0.5 × 19 = 9.5
		3	4.5	1.286		0.3 × 19 = 5.7
		4	5.5	1.57		0.17 × 19 = 3.23

Table 9.3. *Estimation of the values of the magnetic fields H_d for the EMVCo antenna*

We start again and then measure the voltages at different positions of the "barrel" of the EMVCo array, firstly on the EMV_PICC, and secondly the magnetic field at the same point, using the "ISO reference calibration coil".

9.3.2. *Example 2*

So as to avoid any surprises, here are some measurements of the same type, but with a weaker initial field and different loading effects of the target (Table 9.4).

EMV PCD	EMV PICC	
Field HA/m	EMV positions	DC voltage (V)
8.1	000	6.00
7.4	010	6.01
6.4	013	6.00
7.0	016	5.99
6.6	019	6.00
5.2	100	5.62
4.8	110	5.36
4.4	113	5.13
4.5	116	5.33
4.4	119	5.21
3.9	120	4.76
3.2	123	4.21
3.7	126	4.69
3.3	129	4.41
3.3	200	4.53

Table 9.4. *Example of measurements with a weaker initial field and different loading effects*

9.3.2.1. *A few easy to perform measurements*

Let us take a new example. Put very simply, we attack the EMV_PCD circuit, tuned and adapted with a sine generator with internal impedance of 50 Ω, delivering 20 V_pp in a no-load situation. With the loading circuit being adapted, we now measure a voltage of 10 V_pp at the output of the 50 Ω generator: this voltage is applied to the circuit.

The effective power actually consumed, therefore, is:

$$P = (10/2/\sqrt{2})^2/50 = 250 \text{ mW}$$

We then measure the voltage at the terminals of the resistance of R = 1 Ω expressed in serial in the setup. We find:

$$V = 0.9 \text{ V\_pp}$$

The value of the effective current in that resistance and thus in the antenna is:

$I\_r.m.s. = 0.9/2/\sqrt{2}/1 = 318$ mA

The power dissipated in that resistance R is:

$P = 1 \times 0.318^2 = 101$ mW

As the other components in the setup are purely reactive, there are 250 – 101 = 149 mW missing! The 149 mW power is dissipated in an equivalent resistance arranged in serial with the 1 Ω. In order to dissipate this power, we need to add a resistance Rx in series with the antenna, of the value:

$0.149 = Rx \times 0.318^2$ whether $Rx = 1.47\ \Omega$

Consider a total resistance Rt in a series with the antenna:

$Rt = 1 + 1.47 = 2.47\ \Omega$

Thus, the true quality coefficient Q of the circuit can be calculated, and must be:

$Q = 50.24/2.47 = 20.3$

This is what EMVCo gives.

This is all very well, but where do the missing 1.47 Ω come from? (To illuminate this mystery, see the following).

In our case (with a 20 V_pp no-load generator, which is 10 V adapted to 50 Ω), the field at the center of the antenna is:

$H = 2 \times 0.318/2/0.035 = 9.085$ A/m r.m.s.

Let us now attempt to answer the question: where have the missing 1.47Ω gone?

Remember (see Chapter 8) that it is possible to use a VNA to characterize an antenna, so as to establish its *equivalent serial diagram* La, Ra (in our case, La = 0, 590 μH and Ra = 2.47 Ω) – see Figure 9.7.

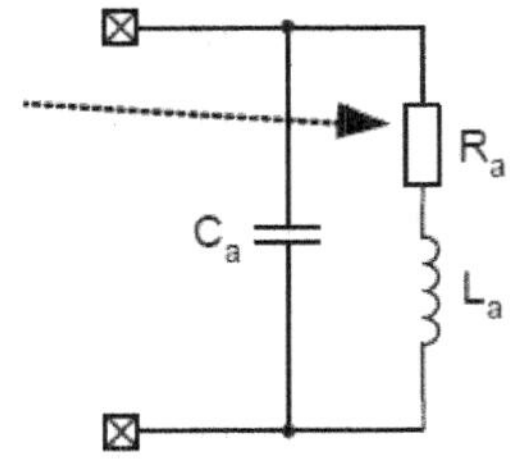

Figure 9.7. *Serial equivalent diagram La, Ra*

However, its *physical diagram* is that of a coil with La = 0.590 μH, of a serial resistance of the coil Rs (at 13 MHz) and a loss resistance (at 13 MHz) arranged in parallel to the coil, as shown in Figure 9.8.

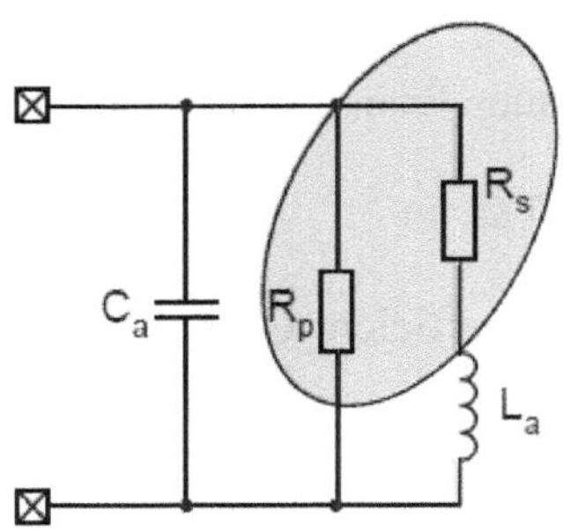

Figure 9.8. *Physical diagram*

In our case, the value of Q is 20.3 at 13.56 MHz, and thus:

Ra@13 MHz = La ω/Q = 50.24/20.3 = 2.47 Ω

If the value Ra@13 MHz is equal to 2.47 Ω, it is because the addition to the sum of the physical resistances which are already in a series on the diagram is equal to Rs = (1 Ω + 0.28 Ω) (on the EMVCo diagram) + 0.22 Ω (R_coil) = 1.5 Ω. We can then calculate Rp@13MHz:

Rp@13 MHz = $(L\ \omega)^2/(Ra - Rs)$

$= (50.24)^2/(2.47 - 1.5)$

= 2602 Ω

and:

$$Rpfra = 2602 \times \sqrt{(13.56/1)}$$

$$= 9581.6\ \Omega$$

At this level, all understanding between the EMV_PCDs and their relations with the values of the magnetic fields H is practically achieved.

9.3.3. *Conformity tests of the NFC Forum*

For its part, the NFC Forum has also developed its own conformity tests for the "Analog" aspects (not yet available to non-members), in the same spirit as those of EMV (measures of voltages rather than fields, distance with the measuring barrels, etc.), slightly relaxing the ISO's constraints in terms of antennas of appropriate forms (three types for PCDs – *polling devices* – and three more for PICCs – *listening devices*; see Figure 9.9), with form factors and field values more closely suited to the movement of the applications intended by the members of the NFC Forum.

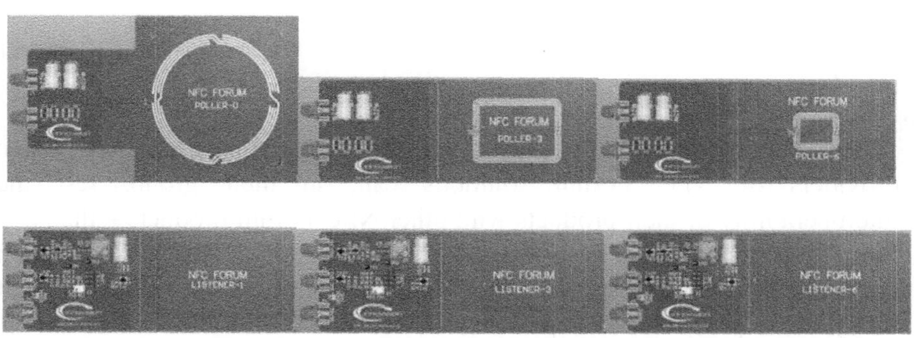

Figure 9.9. *NFC Forum polling devices and listening devices*

9.3.3.1. *Some additional information*

Generally, the antenna of a commercial NFC Forum device has a surface area of around 3 × 3 cm = 900 mm^2 to 3 × 5 cm = 1500 mm^2. This (small) surface area is, notably, a physical constraint necessary for the ergonomics of the device (but can be circumvented if great care is taken) in order to be able to provide remote power to a passive tag or passive card emulator. As the situation currently stands, this constraint has led the NFC Forum to try to reduce the minimum reading distance required for certification in reader mode to around 5 mm (excluding the tolerance margin of a few mm either way), which is too small to allow for the reading of the tag in such a way as to be ergonomic for the end user. For example, a tag placed in a bus shelter under protective glass, or the windscreen/window of a carshare vehicle (respectively 5 and 6 mm in thickness) cannot, in principle, be read by the majority of NFC mobile phones that are simply compliant with the NFC Forum regulations.

Obviously, products which do conform strictly to this standard will be stamped with the "NFC Forum logo", signifying on the market that certain NFC Forum products are compatible with one another, but not necessarily with the NFC IP1 ISO 18 092 or IP2 ISO 21481 norms, and perhaps even less compatible in card-emulation mode with ISO 14443.

Thus, those who choose to remain blinkered as to the physical layers and to these problems of conformity tests between the ISO and the NFC Forum will inevitably encounter serious problems of interoperability with, and access to, markets and ecosystems which do not conform to their views.

9.3.4. *NFC Forum testing and certification methods for the lower levels*

In addition to the "PlugFests" organized several times a year – where providers of NFC Forum solutions can, in their group, test the interoperability of their products with other NFC Forum products – the NFC Forum has established a system of nomenclature and a certification program guaranteeing the end user a certain level of interoperability in the use of NFC Forum products. This provides device manufacturers with a way of establishing consistent behavior between the NFC devices they create, and conformity of their products with the NFC Forum's technical specifications.

9.3.4.1. *Nomenclature*

Before proceeding further, it should be noted that the acronym "NFC" derives from the term "*Near-Field Communication*" and that the term "NFC" is not a registered trademark (nor could it be). That is to say, it is not possible to protect the term itself as a brand name. Nonetheless, the NFC Forum has established two visual identifiers: the "N-Mark" and the "NFC Forum Certified" label.

9.3.4.2. *N-Mark*

The N-Mark logo (see Figure 9.10) is used to identify points which are capable of interacting using an NFC Forum interface.

Figure 9.10. *N-Mark logo*

The position of the logo is supposed to indicate to consumers where to present their NFC device to establish NFC communications and services on a device – e.g. a poster, a badge, a smartphone, etc. Its usual purpose is to mark the place where the NFC device's antenna is located, though it would be unusual to see the N-Mark stuck in the middle of a screen when the device's antenna is arranged all over that screen. Obviously, only businesses whose products have obtained NFC Forum certification may display this sign.

9.3.5. *Testing the conformity of NFC with the CEN – Comité européen de normalisation (European Standardizing Committee)*

For historical reasons, because their industrial day dawned before there was a market for contactless payment, contactless forms of transport are strongly attached to ISO standardization rather than EMVCo. Furthermore, public transport involves specific constraints of quick passage, etc., and the ergonomics of presenting a transport pass (in the form of tickets, cards, mobile phones in card-emulation mode, etc.) to a validator. Thus, it is also necessary for the NFC initiator to be able to provide a certain strength of magnetic field over a certain distance or within a certain volume of space. The values mentioned here are specific to the transport market, and are very different to those involved in the banking and payment markets. Here we see the sectorial nature of the issue.

In order to resolve these issues, on the basis of ISO 14443, at the initiative of the AFIMB – *Agence française pour l'information multimodale et la billettique* (French Multimodal-Information and Ticketing Agency) – the CEN committee TC 278 (Intelligent Transport Systems) constructed a

Europe-wide norm on contactless public transport, ranging from standard chip cards to an NFC mobile phone, whether based on ISO, EMVCo or the NFC Forum, along with tablets and other NFC USB keys. This norm is the European Norm EN TS 16 794-1, and for conformity tests, TS 16 794-2. When establishing these norms, the committee had to deal with all the problems of "gaps" existing between the ISO's norms on chip cards and NFC, the NFC Forum standard and the EMVCo specification, which was no mean feat in terms of the values of the magnetic fields and retromodulated voltage values! Thanks to their work, we are on the road to ensuring a certain degree of interoperability between the ISO, the EMV, and the NFC Forum.

9.3.6. *What about overall interoperability?*

We cannot possibly conclude this chapter without touching on this point.

Following this non-exhaustive overview of all these problems and constraints of interoperability, what conclusion are we to draw?

Interoperability is a delicate problem which is difficult to solve, between applications in different sectors (e.g. banking compared to transport, the automobile domain, the health sector, etc., or all-in-one city applications: transport, school canteens, swimming pools, libraries, assistance to the elderly, etc.) because, owing to the proliferation of the "universal NFC device" (the mobile phone) as one of the main vectors of the proliferation of NFC, its application will be transversal and it will be the sole object used for multi-sector applications. Thus, the technical and commercial interoperability of NFC requires us to overcome all the technical and ergonomic constraints (and their technical consequences) discussed throughout this book.

The CEN initiative discussed above is the first major step in this direction, because it encourages big-name worldwide NFC phone manufacturers to support various norms and standards which tend to be truly interoperable in multi-sector operations (and small mobile manufacturers will surely follow suit). Clearly, all of this cannot be built in a day, but we are on the right path, as indicated at the start of this chapter during our discussion of the interoperabilities of contact chip cards.

Thus, after hours and hours spent at numerous normative and/or standardization committee meetings, we can end this chapter on this optimistic technical note.

Conclusion

The "Near" and "Far" Future of NFC and its Antennas

The technical future of NFC "*as it is*", or almost

Very High Bit Rates (VHBR)

At the time of writing (in 2015), section 2 of the contactless chip-card norm ISO 14443 specifies a basic/initial bit rate of 106 kbps, a High Bit Rate (HBR) of up to 848 kbps, and also supports two versions of Very High Bit Rates (VHBR), with which communication can be extended up to 27 Mbps. The specifications of NFC IP1 and IP2 are limited to 106, 212, and 424 kbps. Bit rates are changing and, in time, if the ISO's NFC norms follow the way of its contactless chip-card norms, the bandwidths and quality factors Q of the antennas are likely to change.

Active Load Modulation (ALM)

Very often, throughout this book, we have mentioned the notion of *active load modulation* (ALM). Indeed, in 2015, for battery-assisted applications (smartphones, watches, bracelets, wearable technology, etc.), the principle of retromodulation from the target to the initiator by *passive load modulation* (PLM) is in the process of being replaced by ALM. In these particular applications, ALM is gaining strength, for two main reasons: firstly, the significant loading effects to which NFC devices are subject (because of their cases, their batteries, etc.) and, secondly, the reduction of the dimensions of antennas (μSD cards, SIM cards, bracelets, etc.) and the associated coupling values.

Reduction of H_min and "low power PCD"

For many years, people have remarked in the field that many (plastic) contactless chip cards are more sensitive than the minimum threshold field of H_min = 1.5 A/m. In addition, mobile phones and other NFC initiators (or, in terms of chip card, PCDs), wishing to reduce their energy consumption for reasons of battery life, are having a great deal of difficult in reaching this threshold field value H_min = 1.5 A/m, and are applying a great deal of pressure (particularly to the ISO) for the value of the emitted field to be reduced to around 0.8 A/m.

The sentiment behind this desire is admirable but, in the interests of interoperability with the vast range of existing chip cards/targets that do conform fully to ISO 14443, it cannot be satisfied. One solution envisaged for the initiator would be for it to be compelled to first test whether it receives responses at 1.5 A/m with targets present in the field and, after a number of exchanges with those targets – during which they communicate whether or not they are able to work with a lower field – begin to save energy.

Discussions are under way on this subject within the ISO. The greatest issue is solving the problems of interoperability with the existing range of devices.

NFC and HF "RFID"

In RFID which, by definition, pertains to the traceability of items, in HF, at 13.56 MHz for many years now, the legislative framework has been carefully constructed in full:

– a norm (ISO 18000-3 mode 1), inspired heavily by the norm on vicinity contactless chip cards (ISO 15693), is often applied for *item management* when those items are present within a certain radius, on display units so with fairly small reading distances (50–70 cm). The collision management principle used for this norm performs far better than ISO 15693;

– a norm (ISO 18000-3 mode 3) is a derivative of the above norm, having the purpose of rendering an HF system at 13.56 MHz capable of supporting the same EPC Global data coding as is used in UHF by ISO 18000-6C;

– given that NFC IP2 supports ISO 15693, it is a very small step to being able to say that a NFC IP2 mobile element (such as a tablet, a mobile phone, etc.) would not have excessive difficulty in reading ECP Global tags via the ISO 18000-3 mod 3 version. Furthermore, the NFC Forum, via semiconductor manufacturers (ST and NXP), is working on the type-5 tag, also known as "V" for Vicinity, in accordance with ISO 15693.

This is an accurate reflection of the work currently in progress.

Tomorrow and beyond

Yet in our wonderful world, it is not only chip cards and mobile phones which use contactless technology. There are also people who make billions of contactless electronic labels – and tags as well – each year and not specifically restricted to 13.56 MHz.

NFC, "RFID" in HF and in UHF

Why, then, would we put two different labels/tags on the same item (e.g. on the collar of a shirt)? The answer is that one is used for UHF, to track it over a medium/long distance (a few meters) throughout the process of manufacture, shipping and delivery; the other is an HF tag (13.56 MHz), used to manage it at a close distance on display units. If we are able to identify the item over a long distance using UHF and are able to simply use UHF to read it close up as well, then we save on the manufacture of a label. In that case, we simply need to provide the public with pocket readers/interrogators working in UHF to read the label, and they can find out whether the product on show is a counterfeit, its origin, if it is made in fair conditions etc.

Once again, we come back to the solution of a pocket-sized item – say, a mobile phone – which must read and communicate, this time in UHF over a short distance, in "near field" or "medium field". From this to speaking of "NFC, or nearly, in UHF", there is only a small step, which some Far-Eastern manufacturers have already taken, some time ago.

RFID, UHF and NFC

Since 2007, the South Koreans at the ETRI – Electronics and Telecommunications Research Institute – have been working on the problem and reached a number of conclusions:

– on the market, at the level of the "items", sooner or later, there will be a proliferation of UHF labels for "supply chain management;

– these labels will primarily be compatible with GS1 and EPC global EPC C1 G2 coding – i.e. in accordance with ISO 18000-6C;

– no one will be happy to put a second label at 13.56 MHz on an article simply so that the general public can read it with a widely-available "pocket reader" – e.g. a mobile phone with 13.56 MHz NFC functions in accordance with an IP1 or IP2 norm;

– if someone wishes to read the content of that UHF label, the mobile phone must be capable of reading something of the same type as ISO 18000-6x;

– as we only wish to read these labels at a short distance (8–15 cm) and do not want to adversely affect the battery life of the mobile phones by overconsumption during the data exchanges, it is preferable to work in UHF rather than HF (a few mW in UHF transmission are sufficient to obtain a communication distance of between 10 and 30 cm) and, if possible, the mobile phone's existing antenna must be used;

– hence, we need to allow for UHF RFID operation of an individual "mobile" device (a phone or another kind of product), using near-field communication. Quite simply, these are NFC products. This is absolutely true; there is no mistaking this appellation.

Yet there is one more small detail that must be considered. Today, in South Korea, there is one of the world leaders – firstly in mobile telephony and secondly in integrated circuits. Its name: Samsung! Of course, it is true that 13.56 MHz NFC will continue on its merry way, using contactless chip-card-emulation application wherever this type of cards reigns (payment, transport, access control, etc.), but a "UHF NFC" version is far from unfounded. Thus, if we want to cast off the blinkers of HF NFC, let us take a look at the following paragraphs on UHF NFC: norms on "RFID mobiles".

Norms on "RFID mobiles"

A norm using the conventional UHF RFID frequency bands (860–960 MHz) has been devised for *mobile item identification and management (MIIM)*: namely ISO 29143 (*Information technology, Automatic identification and data capture techniques, Air interface specification for Mobile RFID interrogators)* which constitutes the ISO norm for the physical layer of UHF NFC systems for "RFID mobiles".

This norm sets out all of the specifications of communication protocols of the "air interface" for mobile systems of RFID interrogators in consumer electronics (e.g. a mobile phone, a tablet, etc.), working with "passive/ backscatter" tags or battery-assisted tags, and interrogator talks first (ITF) systems.

As with conventional RFID, the tags are remotely powered by the (electro) magnetic field supplied by the interrogator's RF signal, and respond to it by modulating the reflection coefficient of their antennas – i.e. by "backscattering" the return data to the interrogator. The tags and the tag-to-initiator communication, therefore, are purely passive. Given that several interrogators and several tags may talk simultaneously, the norm describes how to manage carrier frequency collisions and data collisions from the tags. In addition, the norm describes how two interrogators can communicate with one another.

Convergences between HF and UHF NFC

In terms of applications, it is difficult to imagine:

– firstly, end users having two different mobiles: one for transport and banking applications and another to read product labels (obtaining information about the products' origin, mode of use, warranty, etc.);

– secondly, having two totally-distinct architectures for the functions of HF and UHF NFC.

This leads directly to examine, in time, a possible convergence of NFC solutions which must operate simultaneously in HF and UHF. There may well be a degree of scepticism about this prospect, but it should be noted that in 2007, Samsung created prototypes of the UHF NFC system, and that Nokia also looked into the subject in 2008, presenting a prototype of the concept "HF and UHF NFC" using semi-discrete components. Obviously, as indicated, it is helpful to have a single mobile containing both NFC entities – "NFC HF 13.56 MHz" and "NFC UHF 900 MHz". Obviously, it somewhat complicates the manufacturing process when the device uses discrete components, because it is costly and consumes a little too much energy. However, if the system used is integral, it is entirely achievable.

Common reception head for HF and UHF

Let us examine the technical and technological problems posed by the integration of an RF head shared between the HF and UHF NFC solutions briefly, and the possible ways of solving these problems.

In the reception and decoding apparatus

The channel for reception and demodulation of the multi-band signal (HF to UHF – 10 to 1000 MHz) can be created with *software-defined radio* (SDR) architecture to decode and interpret the NFC IP1 and IP2 norms, ISO 18000 – 3 mod1 and mod3 with 13.56 MHz and ISO 18000 – 6 C, compatible with EPC C1 G2 and ISO 29143 – RFID mobile in UHF. In addition, conventional CMOS technology is widely used to make RF heads (amplification, mixer, FI, zero FI, etc.), either HF or UHF.

The main problem needing to be solved lies in the design of the architecture of an *HF – UHF mono-head*, for an acceptable price.

At the level of the antenna

The ultimate end is to create an *HF and UHF mono-antenna* (gain, impedance matching, radiation pattern, etc.) and to be able to work simultaneously in HF and UHF, in near and far fields.

In order to do so, an *HF/UHF mono-chip* is connected to an appropriate combination of a dipole (for the radiation of far fields in UHF) and a loop (for the inductive coupling of near fields in HF and UHF).

An example of a *mono-antenna* is shown in Figure C.1.

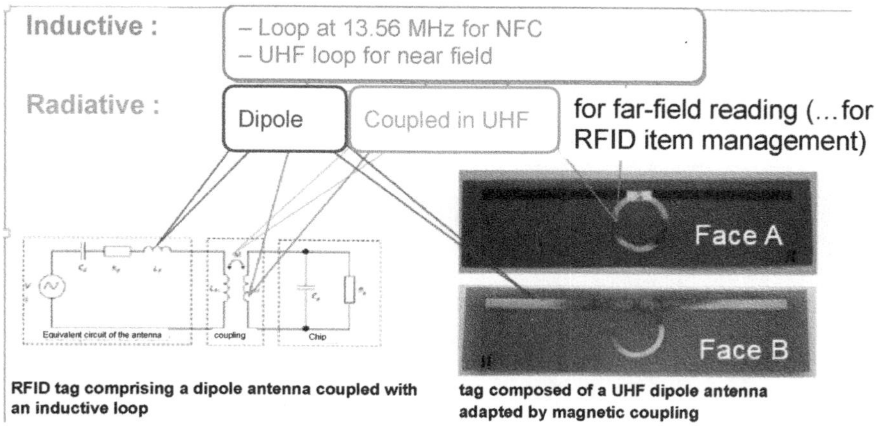

Figure C.1. *Example of an HF and UHF mono-antenna. For a color version of the figure, see www.iste.co.uk/paret/antenna.zip*

There are many other possible solutions out of all distinguished antenna manufacturers on the planet, to which readers should refer for further details.

In addition to all of this, we use a *cryptographic mono-unit* (if possible, of small dimensions, made of silicom and inexpensive – e.g. AES 128 or similar), and this leads back to Chapter 6 for applications of tags linked to the struggle against counterfeiting, with T2-type NFC tags working in HF and UHF.

There are many technical wonders still to come but, in this field of activity, it should absolutely not be forgotten that we must always work on the basis of the principle that, if we do not do something ourselves, someone else will do it instead, and therefore, sooner or later, the solution will come into being.

To push the boat out, there is every possibility that, in a few years' time, we will see NFC devices working in HF and UHF with a mono-head, mono-chip, mono-antenna, and mono-crypto, also usable for Internet of Things applications?

Perhaps this could be the subject of the author's next book. Who knows?

Bibliography

Author bibliography

Glossary
ISO 19 762-3 – standardized vocabulary – radiofrequency identification (RFID).

Norms – Contactless chip-free cards
ISO 14 443 – 1 to 6 – 2015 edition – Proximity cards.
ISO 15 693 – 1 to 3 – 2015 edition –Vicinity cards.
JIS X6319 – 4 – High speed proximity cards – FeliCa.

Norms – NFC
ECMA 340 & 352.
ISO 18 092 – NCF IP 1.
ISO 21 481 – NFC IP2.

Norms – RFID
ISO 18 000 – 3 – 6C and – 63 – RFID, *Supply Chain Management.*

Norms – RFID Mobiles in UHF
ISO 29 143.

Norms – Contactless Communication for Public Transport
CEN 16 794 – 1 protocol.

Norms – Tests
ISO 10 373 – 6 – edition 2015 – conformity tests for proximity cards.
ISO 22 536 – conformity tests for NFC IP1 RF interface.
ISO 23 917 – conformity tests for NFC IP1 protocol.
CEN 16 794 – 2 conformity tests, public transport, contactless.

Norms – Privacy
Mandate M436 of the European Commission.
CEN EN 16 571 – *Privacy Impact Assessment* (PIA).

Norms – Marking of electrical and electronic equipment
EN 50419.

Proprietary standards
EMVCo L1 – contactless bank cards.
NFC Forum – Analog & Digital.

Regulations
ERC 70 03 – relating to the use of short range devices (SRDs).
ETSI 300 330 – conformity tests for radiofrequencies.
US code of federal regulations (CFR) Title 47, Chapter 1, Part 15, "Radio Frequency Devices".

"Human exposure" recommendations
ICNIRP International Commission on Non Ionizing Radiation Protection.

Publisher bibliography

[COK 12] COSKUN V., OK K., OZDENIZCI B., *Near Field Communication (NFC): From Theory to Practice,* Wiley, New York, 2012.

[LAH 14] LAHEURTE J.M., RIPOLL C., PARET D. *et al.*, *UHF RFID Technologies for Identification and Traceability*, ISTE, London and John Wiley & Sons, New York, 2014.

[PAR 03] PARET D., *Applications en identification radiofréquence et cartes à puces sans contact*, Dunod, Paris, vol. 1, 2003.

[PAR 05] PARET D., *RFID and Contactless Smart Card Applications*, Wiley, New York, 2005.

[PAR 12] PARET D., *NFC (Near Field Communication). Principes et applications de la communication en champ proche*, Dunod, Paris, 2012.

[PAR 16] PARET D., *Antenna Designs for NFC Devices*, ISTE, London and John Wiley & Sons, New York, 2016.

Index

J, L, M

N, P, R

S, T, V

Other titles from

iSTE

in

Waves

2016

ANSELMET Fabien, MATTEI Pierre-Olivier
Acoustics, Aeroacoustics and Vibrations

PARET Dominique
Antennas Designs for NFC Devices

WIART Joe
Radio-Frequency Human Exposure Assessment

2015

PICART Pascal
New Techniques in Digital Holography

2014

APPRIOU Alain
Uncertainty Theories and Multisensor Data Fusion

JARRY Pierre, BENEAT Jacques N.
RF and Microwave Electromagnetism

LAHEURTE Jean-Marc
UHF RFID Technologies for Identification and Traceability

SAVAUX Vincent, LOUËT Yves
MMSE-based Algorithm for Joint Signal Detection, Channel and Noise Variance Estimation for OFDM Systems

THOMAS Jean-Hugh, YAAKOUBI Nourdin
New Sensors and Processing Chain

TING Michael
Molecular Imaging in Nano MRI

VALIÈRE Jean-Christophe
Acoustic Particle Velocity Measurements using Laser: Principles, Signal Processing and Applications

VANBÉSIEN Olivier, CENTENO Emmanuel
Dispersion Engineering for Integrated Nanophotonics

2013

BENMAMMAR Badr, AMRAOUI Asma
Radio Resource Allocation and Dynamic Spectrum Access

BOURLIER Christophe, PINEL Nicolas, KUBICKÉ Gildas
Method of Moments for 2D Scattering Problems: Basic Concepts and Applications

GOURE Jean-Pierre
Optics in Instruments: Applications in Biology and Medicine

LAZAROV Andon, KOSTADINOV Todor Pavlov
Bistatic SAR/GISAR/FISAR Theory Algorithms and Program Implementation

LHEURETTE Eric
Metamaterials and Wave Control

PINEL Nicolas, BOURLIER Christophe
Electromagnetic Wave Scattering from Random Rough Surfaces: Asymptotic models

SHINOHARA Naoki
Wireless Power Transfer via Radiowaves

TERRE Michel, PISCHELLA Mylène, VIVIER Emmanuelle
Wireless Telecommunication Systems

2012

LALAUZE René
Chemical Sensors and Biosensors

LE MENN Marc
Instrumentation and Metrology in Oceanography

LI Jun-chang, PICART Pascal
Digital Holography

2011

BECHERRAWY Tamer
Mechanical and Electromagnetic Vibrations and Waves

GROUS Ammar
Applied Metrology for Manufacturing Engineering

GOURE Jean-Pierre
Optics in Instruments

LE CHEVALIER François, LESSELIER Dominique, STARAJ Robert
Non-standard Antennas

2010

BEGAUD Xavier
Ultra Wide Band Antennas

MARAGE Jean-Paul, MORI Yvon
Sonar and Underwater Acoustics

2009

BOUDRIOUA Azzedine
Photonic Waveguides

BRUNEAU Michel, POTEL Catherine
Materials and Acoustics Handbook

DE FORNEL Frederique, FAVENNEC Pierre-Noël
Measurements using Optic and RF Waves

FRENCH COLLEGE OF METROLOGY
Transverse Disciplines in Metrology

2008

FILIPPI Paul J.T.
Vibrations and Acoustic Radiation of Thin Structures

LALAUZE René
Physical Chemistry of Solid-Gas Interfaces

2007

KUNDU Tribikram
Advanced Ultrasonic Methods for Material and Structure Inspection

PLACKO Dominique
Fundamentals of Instrumentation and Measurement

RIPKA Pavel, TIPEK Alois
Modern Sensors Handbook

2006

BALAGEAS Daniel *et al.*
Structural Health Monitoring

BOUCHET Olivier *et al.*
Free-Space Optics

BRUNEAU Michel, SCELO Thomas
Fundamentals of Acoustics

FRENCH COLLEGE OF METROLOGY
Metrology in Industry

GUILLAUME Philippe
Music and Acoustics

GUYADER Jean-Louis
Vibration in Continuous Media

Printed and bound by CPI Group (UK) Ltd, Croydon, CR0 4YY
13/09/2026
14961021-0001